Lecture Notes in Mathematics

A collection of informal reports and seminars
Edited by A. Dold, Heidelberg and B. Eckmann, Zürich

108

Algebraic K-Theory and its Geometric Applications

Edited by R. M. F. Moss and C. B. Thomas
University of Hull

Springer-Verlag
Berlin · Heidelberg · New York 1969

Library of Congress Catalog Card Number 74-97991 Printed in Germany. Title No. 3264

PREFACE

These notes are the texts of lectures given at a conference on "Algebraic K-theory and its Geometric Applications", held at the University of Hull between 17th and 21st March, 1969.

With one exception the texts have been prepared by the various authors, and only marginally revised by the editors. Special thanks are due to Mrs. Margaret Saunders and Mrs. Christine Jones for the speed and accuracy with which they have produced the final typescript, in spite of the pressures of the summer term. Any errors which remain are of course ours.

R. M. F. Moss

C. B. Thomas

Hull, July 1969

CONTENTS

K_2 AND SYMBOLS

by HYMAN BASS

Milnor [3] has introduced a functor K_2 on rings whose definition and basic properties are discussed below. Following that I shall describe some calculations of K_2 for global fields due principally to C. Moore and J. Tate. An exposition of this material is now in preparation by Tate and myself.

1. Homology of groups

Let G be a group, and write $H_i(G) = H_i(G,Z)$, where G acts trivially on Z. For any G-module C with trivial G-action the universal coefficient theorem asserts the existence of an exact sequence

$$0 \to \mathrm{Ext}^1_Z(H_{n-1}(G),C) \to H^n(G,C) \to \mathrm{Hom}_Z(H_n(G),C) \to 0$$

for all $n \geqslant 1$. In particular, if C is Z-injective (e.g. $C = Q/Z$) we have $H^n(G,C) \cong \mathrm{Hom}\,(H_n(G),C)$.

Suppose now that $H_1(G) = 0$, i.e. that G coincides with its commutator group (G,G). Then $H^2(G,C) = \mathrm{Hom}_Z(H_2(G),C)$ is a representable functor of the abelian group C. In particular there is a universal element in $H^2(G,H_2(G))$, corresponding to the identity endomorphism of $H_2(G)$, which is represented by a central group extension

(e) $\quad 1 \to H_2(G) \to \tilde{G} \overset{p}{\to} G \to 1.$

The universality of this extension is expressed by the fact that it maps uniquely into any other central extension

(e') $\quad 1 \to C \to G' \overset{p'}{\to} G \to 1$

of G, i.e. there is a unique homomorphism $h : \tilde{G} \to G'$ such that $p' \circ h = p$. For any homomorphism $d : \tilde{G} \to C$, the map $h'(x) = h(x)\,d(x)$ is a homomorphism such that $p' \circ h' = p$, so $h' = h$, i.e. $d = 1$, by uniqueness. It follows easily, by varying (e'), that $H_1(\tilde{G}) = 0$.

If $g \in \tilde{G}$ and $p(g) \in$ center (G) then $x \mapsto g\,x\,g^{-1}$ defines an automorphism of the extension (e), so is the identity by uniqueness. Hence $g \in$ center ($\tilde{G}$). Suppose $1 \to D \to H \xrightarrow{q} \tilde{G} \to 1$ is a central extension of $\tilde{G}$. The conclusion above implies that $1 \to q^{-1}(C) \to H \xrightarrow{p \circ q} G \to 1$ is a central extension of G, and hence there is a homomorphism $h : \tilde{G} \to H$ such that $(p \circ q) \circ h = p$. Then $q \circ h$ defines an automorphism of (e), hence equals $1_{\tilde{G}}$, so all central extensions of $\tilde{G}$ split, i.e. $H^2(\tilde{G},C) = 0$ for all C. Taking $C = \mathbf{Q}/\mathbf{Z}$ we see that $H_2(\tilde{G}) = 0$.

In summary then, <u>a group</u> G <u>with</u> $H_1(G) = 0$ <u>possesses a universal central extension</u> (e) <u>with</u> $H_1(\tilde{G}) = 0$ <u>and</u> $H_2(\tilde{G}) = 0$. This construction goes back to Schur who called $H_2(G)$ the "multiplicator" of G.

Let G be any group and suppose $\alpha, \beta \in G$ are two elements that commute, i.e. $(\alpha,\beta) = 1$. They define a homomorphism $f : \mathbf{Z}^2 \to G$ which induces $f_* : H_2(\mathbf{Z}^2) \to H_2(G)$. Now $H_*(\mathbf{Z}^2)$ is an exterior algebra on $H_1(\mathbf{Z}^2) = \mathbf{Z}^2$ so $H_2(\mathbf{Z}^2)$ is generated by the wedge w of the basis elements of $\mathbf{Z}^2$ which map to α and β respectively. We put

$$\alpha * \beta = f_*(w) \in H_2(G).$$

This is an antisymmetric function of the pair (α,β)

which is linear as either variable ranges over the centralizer of the other (held fixed). In case $H_1(G) = 0$ we can compute $\alpha * \beta$ from the universal central extension as

$$\alpha * \beta = (\tilde{\alpha}, \tilde{\beta})$$

where the right side is the commutator of liftings $\tilde{\alpha}$ and $\tilde{\beta}$ of α and β, respectively, to $\tilde{G}$.

2. The functors K_1 and K_2

Let A be a ring, and let

$$GL(A) = \bigcup_{n \geq 1} GL_n(A)$$

be the infinite general linear group over A. Here $GL_n(A)$ is embedded into $GL_{n+m}(A)$ by $\alpha \mapsto \begin{pmatrix} \alpha & 0 \\ 0 & I_m \end{pmatrix}$. We put

$K_1(A) = H_1(GL(A)) = GL(A)/E(A)$,

where $E(A) = (GL(A), GL(A))$, and $K_2(A) = H_2(E(A))$.

In fact $H_1(E(A)) = 0$, so we have a canonical exact sequence (see No.1)

$0 \to K_2(A) \to \widetilde{E(A)} \to GL(A) \to K_1(A) \to 0$

deduced from the universal central extension of E(A).

If A is commutative then the determinant induces a split exact sequence

$0 \to SK_1(A) \to K_1(A) \xrightarrow{\det} A^* \to 0$,

where A^* is the group of units of A and $SK_1(A) = SL(A)/E(A)$.

If A is local then $SK_1(A) = 0$.

K_2 appears rather difficult to compute in general, though the results described below suggest that it contains a great deal of interesting information. The definition of K_2 given by Milnor is based on an explicit construction of $\widetilde{E}(A)$ using earlier work of Steinberg. It was proved by Kervaire and by Steinberg (cf. Swan [7]) that the definition of K_2 above agrees with Milnor's.

Products

Milnor has also introduced a multiplicative structure which is quite fundamental for the sequel. It is defined in several steps, as follows:

Step 1. If $\alpha, \beta \in E(A)$ commute then we have $\alpha * \beta \in H_2(E(A)) = K_2(A)$ (cf. No.1).

Step 2. Suppose $\alpha, \beta \in GL_n(A)$ commute, Then so also do $h_{12}(\alpha) = \mathrm{diag}(\alpha, \alpha^{-1}, I)$ and $h_{13}(\beta) = \mathrm{diag}(\beta, I, \beta^{-1})$ in $GL_{3n}(A)$, and these are actually elements of $E(A)$. Hence we can define

$$\alpha \bigstar \beta = h_{12}(\alpha) * h_{13}(\beta) \in K_2(A).$$

This is a bimultiplicative antisymmetric function of the commuting pair α, β.

Step 3. Suppose A is commutative. Then if $\alpha \in GL_n(A)$ and $\beta \in GL_m(A)$ the elements $\alpha \otimes I_m$ and $I_n \otimes \beta$ of $GL_{nm}(A)$ commute, so we can define $\{\alpha, \beta\} = (\alpha \otimes I_m) \bigstar (I_n \otimes \beta) \in K_2(A)$. This pairing $GL_n(A) \times GL_m(A) \to K_2(A)$ stabilizes properly as n and m tend to infinity, so we obtain finally a pairing $\{\ ,\ \} : K_1(A) \times K_1(A) \to K_2(A)$ which is bilinear and antisymmetric.

Functoriality

Let $f : A \to B$ be a ring homomorphism. It induces a homomorphism $GL(A) \to GL(B)$ and hence homomorphisms $f_* : K_i(A) \to K_i(B)$ $(i = 1, 2)$. In case B is a finitely generated projective (right) A-module one can also define natural homomorphisms $f^* : K_i(B) \to K_i(A)$. For example if B is A-free of rank d then $f^* \circ f_*$ is multiplication by d. Moreover, if A and B are commutative then we have the following important formula for $a \in K_1(A)$ and $b \in K_1(B)$:

$$f^*\{f_*a, b\}_B = \{a, f^*b\}_A.$$

EXAMPLE

Suppose A and B are commutative local rings with B free over A. Then $K_1(A) = A^*$ and $K_1(B) = B^*$. Moreover $f^* : K_1B \to K_1A$ is just the norm $N_{B/A} : B^* \to A^*$. For this reason we sometimes call $f^* : K_2B \to K_2A$ the norm also.

3. Symbols

Let A be a <u>commutative</u> ring, with group of units A^*.

DEFINITION

A <u>symbol</u> on A with values in an abelian group C is a function

$$(\ , \ ) : A^* \times A^* \to C$$

which satisfies the following conditions:

1° (,) is bimultiplicative.

2° If a, $1-a \in A^*$ then $(a, 1-a) = 1$.

3° If $a \in A^*$ then $(a, -a) = 1$.

4° (,) is antisymmetric.

Actually these conditions are redundant. For if $a, b \in A^*$ it follows from 1° that

$$(ab, -ab) = (a, -a)(a, b)(b, a)(b, -b)$$

so that 1° and 3° imply 4°. Moreover if $a, 1-a \in A^*$ then 1° implies

$$(a^{-1}, 1-a^{-1}) = (a, 1-a)^{-1}(a, -a),$$

so that 1° and 2° imply that $(a, -a) = 1$ whenever $a, 1-a \in A^*$. If $a \in A^*$ and if A is a field then $1-a \in A^*$ unless $a = 1$, so we conclude that: If A is a field then 1° and 2° imply 3°. We shall be interested in symbols mainly in the case when A is a field.

EXAMPLES

1. The <u>tame symbol</u> $(\ , \ )_v$ associated with a discrete valuation v on a field F with residue class field k(v):

$$(a, b)_v \equiv (-1)^{v(a)v(b)} \frac{a^{v(b)}}{b^{v(a)}} \mod \mathfrak{P}_v.$$

$$(\ , \ )_v : F^* \times F^* \to k(v)^*.$$

It is a simple exercise to verify that $(\ , \ )_v$ is indeed a symbol.

2. The <u>real symbol</u>

$$(\ , \ )_\infty : R^* \times R^* \to \mu_\infty = \{\pm 1\}$$

$$(a, b)_\infty = \begin{cases} -1 & \text{if } a, b < 0 \\ 1 & \text{otherwise.} \end{cases}$$

3. The <u>universal symbol</u>

$$(\, , \,) : A^* \times A^* \to \mathrm{Sym}(A)$$

is the evident symbol with values in the abelian group Sym(A) with generators $A^* \times A^*$ and relations dictated by 1°, ..., 4°. Thus a symbol on A with values in a group C is equivalent to a homomorphism $\mathrm{Sym}(A) \to C$.

4. <u>The symbol defined by</u> K_2

Let A be a commutative ring. We can view elements $a, b \in A^*$ as elements of $K_1(A)$ and thus form $\{a, b\} \in K_2(A)$ (cf. No.2). The next theorem is essentially a consequence of results in Steinberg [6] (cf. also C. Moore [4] and Matsumoto [2]).

THEOREM

The natural pairing

$$\{ \, , \, \} : A^* \times A^* \to K_2(A)$$

is a symbol; hence it induces a homomorphism $\mathrm{Sym}(A) \to K_2(A)$. *If* A *is a field the latter is surjective*.

In fact it has been shown by M. Stein [5] (cf. also A. Christophedes [1]) that $\mathrm{Sym}(A) \to K_2(A)$ is surjective for any local ring A.

The fundamental result for fields is the following consequence of theorems in the thesis of Matsumoto [2].

THEOREM

(Matsumoto). *If* F *is a field then* $\mathrm{Sym}(F) \to K_2(F)$ *is an isomorphism*.

5. Some calculations for global fields

Let F be a global field, i.e. either a finite extension of Q or the function field of a curve over a finite field. If v is a place of F we write F_v for the corresponding completion of F, and call v complex if $F_v \cong C$.

Put

$$\mu_v = \begin{cases} \mu(F_v), \text{ the roots of unity in } F_v, \text{ if } v \text{ is not complex.} \\ \{1\} \text{ if } v \text{ is complex.} \end{cases}$$

From local class field theory one obtains the so called Hilbert symbols

$$(\frac{\ ,\ }{v}) : F_v^* \times F_v^* \to \mu_v,$$

and these induce epimorphisms

$$\lambda_v : K_2(F_v) \to \mu_v.$$

For example, $(\frac{\ ,\ }{v})$ is trivial if v is complex, and it is the real symbol $(\ ,\ )_\infty$ if $F_v \cong R$. If v is a finite place and if char(k(v)) does not divide $[\mu_v : 1]$ then $\mu_v \to k(v)^*$ is an isomorphism which permits us to identify $(\frac{\ ,\ }{v})$ with the tame symbol $(\ ,\ )_v$.

THEOREM

(C. Moore [4]). The Hilbert symbol $(\frac{\ ,\ }{v})$ is universal for continuous symbols on F_v with values in locally compact abelian groups.

With the aid of the norm homomorphism on K_2 (cf. No.2) Tate has given a new and much simpler proof of this theorem, with the following refinement: The kernel of λ_v is divisible and it is generated by symbols $\{a, b\}$

where a and b vary over an arbitrarily small neighbourhood of 1 in F_v^*.

The λ_v's altogether define a homomorphism

$$K_2(F) \xrightarrow{\lambda} \Sigma_F = \coprod_v \mu_v .$$

It takes values in the direct sum because, for a, b $\in F^*$, $(\frac{a,b}{v}) = 1$ for almost all v. For $(\frac{a,b}{v}) = (a,b)_v$ for almost all v, and $(a,b)_v = 1$ whenever $v(a) = 0 = v(b)$.

Consider also the homomorphism

$$\Sigma_F \xrightarrow{\rho} \mu(F)$$

where $\rho : \mu_v \to \mu(F)$ is the $(\frac{m_v}{m})$th power map, $m_v = \text{Card } \mu_v$ and $m = \text{Card } \mu(F)$.

It follows from global class field theory that $\rho \circ \lambda = 0$ in the sequence

$$(*) \qquad K_2(F) \xrightarrow{\lambda} \Sigma_F \xrightarrow{\rho} \mu(F) \to 0.$$

In fact class field theory shows further that, after tensoring with $\mathbf{Z}/m\mathbf{Z}$, the sequence (*) becomes exact. But even without tensoring one has the:

THEOREM

(C. Moore [4]). <u>The sequence</u> (*) <u>is exact</u>.

This theorem can be interpreted as a strong uniqueness theorem for the explicit reciprocity law in class field theory. A new proof of this theorem, again using the norm, has been found by Tate.

In order to complete the calculation of $K_2(F)$ one is next led to investigate $\text{Ker}(\lambda)$. So far the principal result is the following:

THEOREM

Ker(λ) _is finitely generated_. _If_ char(F) = p > 0 _then_ Ker(λ) _is even a finite group of order prime to_ p.

Tate has further shown that Ker(λ) = 0 for F = $\mathbb{Q}$, for certain "euclidean" fields, and for the rational function fields $\mathbb{F}_q(t)$. In the general case he has made, on the basis of both theoretical and elaborate computational evidence, some conjectures which assert that Ker(λ) is a finite group of a certain predictable order. For the moment, however, we know of no examples where Ker(λ) is not zero.

Institut des Hautes Études Scientifiques
and Columbia University.

REFERENCES

1. A. Christophedes, Structure and presentations of unimodular groups, Thesis, London University (196).

2. H. Matsumoto, Sur les sous-groupes arithmétiques des groups semi-simples déployés, Ann. Scient. Éc. Norm. Sup., 4ème serie, t.2, 1969, p.1-62.

3. J. Milnor, Algebraic K-theory, mimeographed notes, U.C.L.A., 1968.

4. C. Moore, Group extensions of p-adic and adelic linear groups, Publ. Math. I.H.E.S. No. 39, 1969, p.5-74.

5. M. Stein, Thesis, Columbia Univ. (in preparation).

6. R. Steinberg, Générateurs, relations, et revêtements de groupes algebriques, Colloque sur la theorie des groupes algebriques, Bruxelles, 1962, p.113-127.

7. R.G. Swan, Algebraic K-theory, Lecture notes in Mathematics, 76 (1968), Springer-Verlag.

FOUNDATIONS OF EQUIVARIANT ALGEBRAIC K-THEORY*

By A. FRÖHLICH and C.T.C. WALL

1. Recall of ordinary algebraic K-theory

In this section we recapitulate, in our own notation, the salient facts about the exact sequence of algebraic K-theory. Our account closely follows pp. 7-16 of H. Bass, 'Lectures on topics in algebraic K-theory', Tata Institute, Bombay, 1967; to be referred to hereafter as [B].

We assume familiarity with the notions of category, functor etc. If $\mathcal{C}$ is a category, we regard $\mathcal{C}$ as a set of morphisms, and write $|\mathcal{C}|$ for the corresponding set of objects, We write $\mathcal{C}at$ for the category whose objects are the (small) categories and morphisms are functors between them.

A groupoid is a category in which all morphisms are isomorphisms. A category with product is a groupoid $\mathcal{C}$, together with a 'product' functor

$$\oplus : \mathcal{C} \times \mathcal{C} \to \mathcal{C}$$

and natural equivalences

$$A : \oplus \circ (1_{\mathcal{C}} \times \oplus) \cong \oplus \circ (\oplus \times 1_{\mathcal{C}}) : \mathcal{C} \times \mathcal{C} \times \mathcal{C} \to \mathcal{C}$$

$$C : \oplus \circ T \cong \oplus : \mathcal{C} \times \mathcal{C} \to \mathcal{C}$$

(where T is the transposition on $\mathcal{C} \times \mathcal{C}$) which are coherent in that isomorphisms of products of several factors obtained from these by re-bracketing (using the first) and permuting

---

* An account of this work was presented by Wall as the first half of his talk at the conference. The topic discussed in the second half is not yet prepared for publication, and details will appear elsewhere.

the factors (using the second) depend only on the re-bracketing and permutation involved, and not on the particular chain of equivalences chosen.

Although this suffices for K-theory, it seems to us more natural (and for other purposes essential) to postulate also a 'unit' for the product, given by an object 0 of $\mathscr{C}$ and a natural isomorphism $0 \oplus M \cong M$ for objects M of $\mathscr{C}$, coherent with the above. Sometimes, also it is desirable to have an inverse. We also draw attention to the problem of replacing $\oplus$ by a product which is associative and commutative 'on the nose' (i.e. without need for natural equivalences): this will be discussed in detail in our next publication.

Denote by $\mathfrak{P}$ the category whose objects are the (small) categories-with-product $(\mathscr{C}, \oplus, A, C)$, and whose morphisms are the functors $F : \mathscr{C} \to \mathscr{C}'$ together with a natural isomorphism

$$f : F \circ \oplus \cong \oplus' \circ (F \times F) : \mathscr{C} \times \mathscr{C} \to \mathscr{C}'$$

such that for $L, M, N \in |\mathscr{C}|$,

$$C'(FL, FM) \circ f(M, L) = f(L, M) \circ FC(L, M),$$

$$A'(FL, FM, FN) \circ (1_{FL} \oplus' f(M, N)) \circ f(L, M \oplus N) =$$
$$(f(L, M) \oplus' 1_{FN}) \circ f(L \oplus M, N) \circ FA(L, M, N).$$

For each $(\mathscr{C}, \oplus, A, C) \in |\mathfrak{P}|$, the abelian group $K_0(\mathscr{C})$ and map $(\ )_{\mathscr{C}} : |\mathscr{C}| \to K_0(\mathscr{C})$ are universal subject to:

(a) If $M \cong N$, then $(M)_{\mathscr{C}} = (N)_{\mathscr{C}}$

(b) $(M \oplus N)_{\mathscr{C}} = (M)_{\mathscr{C}} + (N)_{\mathscr{C}}$.

It is easily seen that we have a functor $K_0 : \mathfrak{P} \to \mathcal{A}b$, where

$\mathcal{A}b$ is the category of abelian groups.

Next, a composition on a category $\mathcal{C}$ with product is a sometimes defined composition $\circ$ of objects of $\mathcal{C}$ such that

$$(M \circ M') \oplus (N \circ N') = (M \oplus N) \circ (M' \oplus N')$$

in the sense that if the left hand side is defined, so is the right, and the two are equal. Write $\mathcal{P}^c$ for the category whose objects are categories with product and composition and morphisms are morphisms in $\mathcal{P}$ which satisfy

$$F(M \circ N) = FM \circ FN.$$

We define a functor $K_0^c : \mathcal{P}^c \to \mathcal{A}b$ as above, but requiring in addition to (a) and (b)

(c) $(M \circ N)_{\mathcal{C}} = (M)_{\mathcal{C}} + (N)_{\mathcal{C}}$.

The loop and fibre functors

$$\Omega : \mathcal{P} \to \mathcal{P}^c \qquad \Phi : \mathcal{P}^2 \to \mathcal{P}^c$$

are defined as follows. For $(\mathcal{C}, \oplus, A, C) \in |\mathcal{P}|$, the objects of $\Omega\mathcal{C}$ are the automorphisms $\alpha : M \to M$ of objects of $\mathcal{C}$ which we shall also write as pairs (M, α). A product functor, with associating and permuting natural equivalences, are then induced from $\oplus$, A and C respectively; and we define composition to be composition of morphisms in $\mathcal{C}$. Similarly, given a morphism in $\mathcal{P}$,

$$(F, f) : (\mathcal{C}, \oplus, A, C) \to (\mathcal{C}', \oplus', A', C'),$$

the objects of ΦF are the triples (M, N, α) with $M, N \in |\mathcal{C}|$ and $\alpha : FM \to FN$ an equivalence; a morphism $(M_1, N_1, \alpha_1) \to (M_2, N_2, \alpha_2)$ is a pair (β, γ) with $\beta : M_1 \to M_2$, $\gamma : N_1 \to N_2$ isomorphisms in $\mathcal{C}$ such that $\alpha_2 \circ F\beta = F\gamma \circ \alpha_1$.

We define a product by

$$(M_1, N_1, \alpha_1) \oplus (M_2, N_2, \alpha_2) = (M_1 \oplus M_2, N_1 \oplus N_2, f(N_1, N_2)^{-1} \circ (\alpha_1 \oplus' \alpha_2) \circ f(M_1, M_2)),$$

A and C then induce natural equivalences in an obvious way. Finally, the composition of (M_1, N_1, α_1) and (M_2, N_2, α_2) is defined just when $N_1 = M_2$ and then is $(M_1, N_2, \alpha_2 \circ \alpha_1)$.

Although we have only described Ω and Φ on objects, it is easy to extend them to morphisms, and obtain functors. The composite $K_0^c \circ \Omega : \mathcal{P} \to \mathcal{A}b$ is denoted by K_1.

For the exact sequence, one further concept is needed. A functor in $\mathcal{P}$, $F : \mathcal{C} \to \mathcal{D}$, is called cofinal if for each $N \in |\mathcal{D}|$ we can find $N' \in |\mathcal{D}|$ and $M \in |\mathcal{C}|$ with $F(M)$ isomorphic to $N \oplus N'$. If these choices can be made naturally, i.e. if there are functors $H : \mathcal{D} \to \mathcal{C}$, $J : \mathcal{D} \to \mathcal{D}$ and a natural equivalence of $F \circ H$ on $1_{\mathcal{D}} \oplus J$, we call F naturally cofinal: this is a more useful property but excludes (as we will see below) some important examples of cofinal functors.

THEOREM 1. (The exact sequence of algebraic K-theory)[B]. _If_ $F : \mathcal{C} \to \mathcal{D}$ _is a cofinal functor in_ $\mathcal{P}$, _there is an exact sequence_

$$S(F) : K_1(\mathcal{C}) \xrightarrow{K_1(F)} K_1(\mathcal{D}) \xrightarrow{U(F)} K_0^c\Phi(F) \xrightarrow{V(F)} K_0(\mathcal{C}) \xrightarrow{K_0(F)} K_0(\mathcal{D})$$

Remarks

$V(F)$ is induced by taking the class of $(M, N, \alpha) \in |\Phi(F)|$ to $(M)_{\mathcal{C}} - (N)_{\mathcal{C}}$; exactness at $K_0(\mathcal{C})$ is relatively easy to prove. If F is naturally cofinal, we can define a functor $G \in \mathfrak{P}^c$, $G : \Omega(D) \to \Phi(F)$ by

$$G(N, \alpha) = (H(N), H(N), g_N^{-1}(\alpha \oplus 1_{J(N)})g_N)$$

where $g_N : FH(N) \to N \oplus J(N)$ is the given natural isomorphism. We then have $U(F) = K_0^c(G)$. The definition in general is essentially the same.

2. G-graded categories and the equivariant groups.

We next discuss graded categories. Let G be a group. We regard G as a category with only one object (and all morphisms invertible). A G-graded category (or category over G) is a pair $(\mathscr{C}, \gamma)$ with $\mathscr{C}$ a category and $\gamma : \mathscr{C} \to G$ a functor. Write $\mathscr{C}at_G$ for the category whose objects are (small) G-graded categories and morphisms $(\mathscr{C}, \gamma) \to (\mathscr{C}', \gamma')$ are functors $F : \mathscr{C} \to \mathscr{C}'$ with $\gamma' \circ F = \gamma$.

This category can be enriched: we define a category $\mathscr{H}om((\mathscr{C}, \gamma), (\mathscr{C}', \gamma'))$ whose objects are the functors F above and whose morphisms $F_1 \to F_2$ are natural transformations $\eta : F_1 \to F_2$ such that for each $C \in |\mathscr{C}|$, $\gamma'(\eta(C)) = 1$. It might seem more natural here to define η of grade $g \in G$ by requiring $\gamma'(\eta(C)) = g$. However, this is easily seen to imply that g is in the centre of G; such η only seem to be important when G is abelian (particularly, free abelian).

Since $\mathscr{C}at$ has pullbacks, $\mathscr{C}at_G$ has products. The pair $(G, 1_G)$ acts as unit for this product. We define $\mathrm{Rep} : \mathscr{C}at_G \to \mathscr{C}at$ by

$$\mathrm{Rep}(\mathscr{C}, \gamma) = \mathscr{H}om((G, 1_G), (\mathscr{C}, \gamma));$$

as a hom functor, this preserves products. The name of the functor arises from the concept of representations of G in the category $\mathscr{C}$. We have generalised the ordinary concept of representation (consider arbitrary functors $G \to \mathscr{C}$) by having the grading as extra structure: note that from any (ungraded) category $\mathscr{D}$ we can construct a G-graded category $(\mathscr{D} \times G, \pi_2)$, (where π_2 is projection on the second factor), and objects of $\mathrm{Rep}(\mathscr{D} \times G, \pi_2)$ are then representations of G in $\mathscr{D}$ in the naïve sense.

We define $\mathrm{Ker} : \mathscr{C}at_G \to \mathscr{C}at$ by restricting to the subcategory of morphisms of grade 1 ($\mathrm{Ker}\ \gamma$); alternatively, we can write

$$\mathrm{Ker}(\mathscr{C}, \gamma) = \mathscr{H}om((1, 1), (\mathscr{C}, \gamma)).$$

For equivariant algebraic K-theory we need the notion of G-graded category with product. If $(\mathscr{C}, \gamma)$ is G-graded, a product is a functor

$$\oplus : \mathscr{C} \times_G \mathscr{C} \to \mathscr{C},$$

where $\times_G$ is the product in $\mathscr{C}at_G$. Thus the sum of two morphisms of $\mathscr{C}$ is only defined when the morphisms have the same grade. We also require natural equivalences as before,

$$A : \mathscr{C} \times_G \mathscr{C} \times_G \mathscr{C} \to \mathscr{C}, \qquad C : \mathscr{C} \times_G \mathscr{C} \to \mathscr{C},$$

which are still to be coherent. Defining morphisms as previously, we construct the category $\mathscr{P}_G$ of G-graded categories-with-product.

If $(\mathscr{C}, \gamma, \oplus, A, C) \in |\mathscr{P}_G|$, and $(\mathscr{D}, \delta)$ is G-graded, then

$\mathcal{H}om((\mathcal{D}, \delta), (\mathcal{C}, \gamma))$ inherits a product and coherent natural equivalences from those in $\mathcal{C}$, and thus defines an object of $\mathcal{P}$. In particular, we can identify $\mathrm{Rep}(\mathcal{C}, \gamma)$ and $\mathrm{Ker}(\mathcal{C}, \gamma)$ with objects of $\mathcal{P}$. We wish to regard $K_0(\mathrm{Ker}(\mathcal{C}, \gamma))$ as a G-module by specifying that if $e : M \to N$ is an isomorphism of grade $g \in G$, then

$$g.(M)_{\mathrm{Ker}(\mathcal{C}, \gamma)} = (N)_{\mathrm{Ker}(\mathcal{C}, \gamma)}.$$

For this we need an extra axiom. Say that $(\mathcal{C}, \gamma) \in |\mathcal{C}at|_G$ is <u>stable</u> if for all $M \in |\mathcal{C}|$, $g \in G$, there exists a $N \in |\mathcal{C}|$ and an isomorphism $e : M \to N$ of grade g (i.e. with $\gamma(e) = g$). We write $\mathcal{SC}at_G$, $\mathcal{SP}_G$ for the full subcategories of $\mathcal{C}at_G$, $\mathcal{P}_G$ whose objects are stable. Notice that N is determined up to an ismorphism of grade 1, i.e. in $\mathrm{Ker}(\mathcal{C}, \gamma)$. It follows that we have defined an operation of G on the set of isomorphism classes of objects of $\mathrm{Ker}(\mathcal{C}, \gamma)$. Since the operation respects products, we inherit an operation on $K_0(\mathrm{Ker}(\mathcal{C}, \gamma))$, which thus becomes a left G-module. We have thus constructed a functor $\mathcal{SP}_G \to {}_G\mathcal{M}od$; by abuse of notation we will denote it by K_0.

The notion of composition on a G-graded category-with-product is defined as in the ungraded case. But we need a further stability axiom to be sure that $K_0^c(\mathrm{Ker}(\mathcal{C}, \gamma))$ inherits a G-module structure, for it is a quotient of K_0 defined by the relations (c): we want G to respect these. Thus an object of $\mathcal{P}_G^c$ will be called stable if whenever $L \circ M$ is defined and $g \in G$, there exist isomorphisms of grade g $L \to L'$, $M \to M'$, $L \circ M \to N'$ for some objects L', M', N' with

$L' \circ M'$ defined, and an isomorphism of grade 1 $L' \circ M' \to N$. We let $\mathcal{SC}_G^c$ be the corresponding full subcategory. If $(\mathscr{C}, \gamma)$ defines an object of this, then (as we have just seen) G operates on $K_0^c(\mathscr{C})$, making it a left G-module; thus we have a functor $K_0^c : \mathcal{SC}_G \to {}_G\mathfrak{Mod}$.

We next define the loop and fibre functors in the graded case. Indeed, the definitions are as above, but with the restriction that $\gamma(\alpha) = 1$ in each case. This restriction only applies to the objects of $\Omega\mathscr{C}$ and ΦF: morphisms can have any grade. If $\mathscr{C}$ is stable, so is $\Omega\mathscr{C}$, for if $\alpha\circ\beta$ is defined, α and β are automorphisms of the same object L of $\mathscr{C}$. If $e : L \to M$ has grade g, it defines morphisms of grade g in ΩC,

$$\alpha \mapsto e\circ\alpha\circ e^{-1}, \quad \beta \mapsto e\circ\beta\circ e^{-1}; \quad \alpha\circ\beta \mapsto e\circ\alpha\circ\beta\circ e^{-1},$$

and $(e\circ\alpha\circ e^{-1})\circ(e\circ\beta\circ e^{-1}) = e\circ\alpha\circ\beta\circ e^{-1}$. Thus $\Omega : \mathcal{SC}_G \to \mathcal{SC}_G^c$. Similarly if $F : \mathscr{C} \to \mathscr{C}'$ is part of a morphism in $\mathscr{C}_G$ and $\mathscr{C}$ is stable, then so is ΦF.

Since the definition of objects of $\Omega\mathscr{C}$, ΦF involved only morphisms of grade 1, Ω and Φ commute with Ker. We assert that they also commute with Rep: let us check this for Ω. For any object of Rep $\Omega(\mathscr{C}, \gamma)$ we pick an object $\alpha : M \to M$ of $\Omega(\mathscr{C}, \gamma)$ and a homomorphism of G, preserving grade, to the group of $\Omega(\mathscr{C}, \gamma)$-automorphisms of α, i.e. the $\mathscr{C}$-automorphisms of M commuting with α. This defines a representation of G on M, hence an object of Rep$(\mathscr{C}, \gamma)$; now since α commuted with the image of G, it defines an automorphism (of grade 1) of this object, hence an object of Ω Rep($\mathscr{C}$, γ). The argument is reversible and gives a bijection of objects; indeed, an isomorphism of categories Rep $\Omega(\mathscr{C}, \gamma) \to \Omega$ Rep($\mathscr{C}$, γ).

The definition of cofinality is the same as in the ungraded case. In fact $F : (\mathscr{C}, \gamma) \to (\mathscr{D}, \delta)$ is a cofinal functor in $\mathscr{P}_G$ if and only if Ker F : Ker($\mathscr{C}$, γ) $\to$ Ker($\mathscr{D}$, δ) is cofinal in $\mathscr{P}$. For a naturally cofinal functor in $\mathscr{P}_G$ one will of course require the functors H and J (in our original definition) to be functors in $\mathscr{P}_G$.

THEOREM 2. (The exact sequence, equivariant version). If $F : (\mathscr{C}, \gamma) \to (\mathscr{D}, \delta)$ is a cofinal functor in $\mathscr{P}_G$, the exact sequence S(Ker F) (see Theorem 1) is one of G-modules. If F is naturally cofinal so is Rep(F). If F and Rep F are cofinal, there is a morphism S(Rep F) $\to$ S(Ker F) of exact sequences, whose image subgroups are element wise fixed by G.

It is clear from the definitions that the maps in S(Ker F) are compatible with the action of G. If F is naturally cofinal, $F \circ H \cong 1_{\mathscr{D}} \oplus J$, then Rep $F \circ$ Rep $H \cong 1_{\text{Rep } \mathscr{D}} \oplus$ Rep J, so Rep F is naturally cofinal. The morphism S(Rep F) $\to$ S(Ker F) comes from ignoring the actions of G: clearly we have a morphism of sequences. It is enough to check the final statement on K_0: but if $M \in |\mathscr{C}|$ underlies a representation of G, it admits automorphisms of all grades, so by definition

$$g.(M)_{\mathscr{C}} = (M)_{\mathscr{C}} \qquad \text{for all } g \in G,$$

as asserted.

3. Change of groups.

The above remarks admit some generalisation when we consider several groups simultaneously. The functor Ker admits an obvious generalisation for any subgroup H of G: forget all morphisms except those whose grade belongs to H. More generally, for any group homomorphism $f : H \to G$ and G-graded category $(\mathscr{C}, \gamma)$ we have an H-graded category. $\mathscr{C} \times_G H$, where H is regarded (via f) as a G-graded category. Note that the objects of this are essentially just those of $\mathscr{C}$. Applying now Rep, we obtain H-representations in $\mathscr{C}$. We write (assuming f known) $\text{Rep}(H, \mathscr{C}) = \text{Rep}(\mathscr{C} \times_G H)$.

Now assume H a normal subgroup of G, $H \triangleleft G$. Then, we claim, $\text{Rep}(H, \mathscr{C})$ can be extended naturally to a G/H-graded category. An object of $\text{Rep}(H, \mathscr{C})$ is a pair (M, φ), where $M \in |\mathscr{C}|$ and $\varphi : H \to \text{Aut}_{\mathscr{C}}(M)$ is a homomorphism such that $\gamma \circ \varphi = f$. A morphism $(M, \varphi) \to (M', \varphi')$ of grade Hg is a set of $\mathscr{C}$-morphisms $\psi(hg) : M \to M'$, with $\gamma(\psi(hg)) = hg$, such that for all $h, h', h'' \in H$, $\varphi'(h'')\circ\psi(h'g)\circ\varphi(h) = \psi(h''h'gh)$. Note it is sufficient to prescribe $\psi_0 = \psi(g)$, of grade g, for then $\psi(hg) = \varphi'(h)\circ\psi_0$. Moreover, this ψ defines a morphism precisely if, for all $h, h', h'' \in H$,

$$\varphi'(h'')\circ\psi(h'g)\circ\varphi(h) = \varphi'(h'')\circ\varphi'(h')\circ\psi_0\circ\varphi(h)$$

and

$$\begin{aligned}\psi(h''h'gh) &= \varphi'(h''h'ghg^{-1})\circ\psi_0 \\ &= \varphi'(h'')\circ\varphi'(h')\circ\varphi'(ghg^{-1})\circ\psi_0\end{aligned}$$

coincide, i.e. iff for all $h \in H$,

$$\psi_0\circ\varphi(h) = \varphi'(ghg^{-1})\circ\psi_0.$$

Now multiplying these sets of morphisms as subsets of $\mathscr{C}$

gives a composition which turns $\mathrm{Rep}(H, \mathscr{C})$ into a G/H-graded category. Also, it is clear that
$\mathrm{Rep}(G/H, \mathrm{Rep}(H, \mathscr{C})) = \mathrm{Rep}(G, \mathscr{C})$.

It is clear that if $\mathscr{C}$ has a product, then so has $\mathscr{C} \times_G H$, regarded as a H-graded category; if $\mathscr{C}$ is stable, so is $\mathscr{C} \times_G H$; similarly for composition. Thus all the above constructions apply to this case.

Considered as depending on H, $\mathrm{Rep}(H, \mathscr{C})$ is in particular a contravariant functor of the category of subgroups of G and inclusion maps, where for $i : H \hookrightarrow K$

$$i^* : \mathrm{Rep}(K, \mathscr{C}) \to \mathrm{Rep}(H, \mathscr{C})$$

is defined by restricting to H the given action of K. On the other hand (provided $(\mathscr{C}, \gamma)$ has a product and $|K : H|$ is finite) we also get a Frobenius induction map.

$$i_* : \mathrm{Rep}(H, \mathscr{C}) \to \mathrm{Rep}(K, \mathscr{C}),$$

defined as follows. Let $\{k_i\}$ be a (say left) transversal of H in K. Given a representation of H, say $\varphi : H \to \mathrm{Aut}_{\mathscr{C}}(M)$, choose isomorphisms $e_i : M \to M_i$ with $j(e_i) = k_i$. One may suppose $k_1 = 1$, $e_1 = 1_M$. Let $N = \bigoplus_i M_i$. We define a representation of K on M. For $k \in K$, let $kk_i = k_{\sigma(i)}h_i$ $(h_i \in H)$. Then $\psi(k)$ is the automorphism of N with constituent isomorphisms $M_i \to M_{\sigma(i)}$ given by $e_{\sigma(i)} \circ \varphi(h_i) \circ e_i^{-1}$. (Note that this definition uses implicitly the natural equivalences A and C). Clearly $\psi : K \to \mathrm{Aut}_{\mathscr{C}}(N)$ is a grade preserving homomorphism. The construction involves some choices but is unique up to equivalence and is clearly seen to be functorial (say on the subgroups of G of finite index).

Composing with K_0 and K_1 we obtain further functors of subgroups of G and inclusion maps, admitting Frobenius induction maps. This property is useful in giving partial results on the $K_i(\text{Rep}\ (G, \mathscr{C}))$ in terms of smaller subgroups of G; in some ways this is the key problem of equivariant algebraic K-theory. Given some further structure, essentially a second product $\otimes$ acting distributively with respect to $\oplus$, K_0 Rep becomes a Frobenius functor in the sense of T.Y. Lam (Induction Theorems for Grothendieck Groups and Whitehead Groups of Finite Groups, Ann.scient. de l'École Normale Sup., t.1, 1968, 91-148). This means that the $K_0(\text{Rep}(H, \mathscr{C}))$ have the structure of commutative rings, the restriction maps $i^* : K_0(\text{Rep}(K, \mathscr{C})) \to K_0(\text{Rep}(H, \mathscr{C}))$ are homomorphisms of rings and the induction maps i_* (cf. Abelian groups) satisfy the Frobenius law

$$i_*(i^*a.b) = a.i_*b$$

(where $i : H \subset K$, $a \in K_0(\text{Rep}(K, \mathscr{C}))$, $b \in K_0(\text{Rep}(H, \mathscr{C}))$.). Similarly K_1 Rep will become a 'module over K_0' in the sense of Lam.

4. Examples.

First we consider the category $\mathfrak{M}_A$ of finitely generated projective A-modules for some ring A, with $\oplus$ as the usual direct sum. Let G operate as a group of automorphisms of A. Then a morphism $M \overset{\varphi}{\to} N$ of grade g is to be a homomorphism of groups such that $\varphi(ma) = \varphi(m)a^g$ for $m \in M$, $a \in A$. Since any projective module is a direct summand of a free module, if $f : A \to B$ is a ring homomorphism, the functor $f_* : \mathfrak{M}_A \to \mathfrak{M}_B$ induced by $M \mapsto M \otimes_A B$ is cofinal, since free modules map to free ones. But f_* is not naturally cofinal, and $\mathrm{Rep}(f_*)$ need not be cofinal.

As an example, let A be the ring Z of integers, G a group $\{1, T\}$ of order 2 operating trivially. Let $B = Z[e | e^2 = 1]$. Modules in $\mathrm{Rep}(\mathfrak{M}_A)$ are well-known to be direct sums of those of the forms

$$Zu (Tu = u), \quad Zv (Tv = -v), \quad Zx \oplus Zy \quad (Tx = y, Ty = x).$$

Forming $\otimes_A B$, we get corresponding modules: observe that in each case, and hence for any sum M of copies of them, the sequence

$$M \xrightarrow{1+eT} M \xrightarrow{1-eT} M \xrightarrow{1+eT} M$$

is exact. However, the free B-module M_e with basis z and $Tz = ez$ does not share this property, so is not a direct summand of any such M. Hence $\mathrm{Rep}(f_*)$ is not cofinal in this case.

It is useful to note that objects of $\mathrm{Rep}(\mathfrak{M}_A)$ are modules over the twisted group ring A[G] whose elements are the (finitely non-zero) sums $\sum_{g \in G} g a_g \quad (a_g \in A)$, with

$$(\Sigma g a_g)(\Sigma h b_h) = \Sigma gh\, a_g^h b_h:$$

in fact we have those A[G]-modules which are finitely generated and projective as A-modules. In some cases this implies them projective as A[G]-modules.

THEOREM. <u>Suppose</u> G <u>finite, and that there exists an element</u> u <u>of the centre of</u> A <u>such that</u> $\Sigma_{g \in G} u^g = 1$. <u>Then</u>

<u>every A-projective</u> A[G]-<u>module</u> M <u>is</u> A[G]-<u>projective</u>.

<u>Proof</u>. Form $M[G] = M \otimes_A A[G]$: the elements may be written as $\Sigma_{g \in G}\, gm_g$ ($m_g \in M$). Let G act 'diagonally', so we have a right A[G]-module structure defined by

$$(\Sigma gm_g)(\Sigma ha_h) = \Sigma gh(m_g^h a_h).$$

We now have A[G]-homomorphisms $p : M[G] \to M$ and $i : M \to M[G]$, with $p \circ i = 1_M$, given by

$$p(\Sigma gm_g) = \Sigma m_g, \quad i(m) = \Sigma gmu^g.$$

Thus M is an A[G]-direct summand of M[G].

It remains to show that M A-projective implies M[G] A[G]-projective. Given an epimorphism $\varepsilon : P \to Q$ of A[G]-modules, and $\varphi : M[G] \to Q$, consider the A-submodule M of M[G] given by $m_g = 0$ for $g \neq 1$. Since M is A-projective, $\varphi|M$ lifts to an A-homomorphism $\psi_0 : M \to P$. We can then define an A[G]-homomorphism ψ lifting φ by

$$\psi(\Sigma gm_g) = \Sigma\, \psi_0(gm_g g^{-1}).g.$$

This result generalises the case when A is commutative and G acts trivially, and the condition is for A[G] to be a separable A-algebra. When the theorem applies, the category $\mathrm{Rep}(\mathfrak{M}_A)$ can be identified with $\mathfrak{M}_{A[G]}$. Note also that if $f : A \to B$ is a ring homomorphism, compatible with with actions of G on A and B, and if B satisfies the condition of the theorem, then $\mathrm{Rep}(f_*)$ is cofinal. For then B[G]-free modules are cofinal in $\mathrm{Rep}(\mathfrak{M}_G)$, and these certainly do lie (up to isomorphism) in the image of f_*.

If G acts trivially on A, the condition is that the

order of G (times the unit of A) be an invertible element of A; when A is a field, this is of course the condition of semi-simplicity. The condition is also fulfilled when A is commutative and G acts as the Galois group of some Galois extension, A over A_0 (the fixed ring of G). In this case, the categories $\mathfrak{M}_A[G]$ and $\mathfrak{M}_{A_0}$ are Morita equivalent; indeed it can also be seen directly that $\mathrm{Rep}(\mathfrak{M}_A)$ is equivalent to $\mathfrak{M}_{A_0}$: the equivalence is induced by looking at the A_0-submodule of elements fixed by G (the other way, form $\otimes_{A_0}$ A, and induce a G-action from that on A). We thus obtain isomorphisms

$$K_i(\mathrm{Rep}(\mathfrak{M}_A)) \cong K_i(\mathfrak{M}_{A_0}) \qquad (i = 0, 1).$$

If A is commutative then we get a second product on the G-graded category $\mathfrak{M}_A$, namely the tensor product $\otimes_A$. The $K_0(\mathrm{Rep}(H, \mathfrak{M}_A))$ for subgroups H of G now have the structure of commutative rings and in fact K_0 Rep becomes a Frobenius functor and K_1 Rep a module over it, in the sense of Lam. The induction map is given by

$$M \longmapsto M \otimes_{A[H]} A[K]$$

and the proof of Frobenius reciprocity is the same as in the well known case when G acts trivially on A (See R.G.Swan, Induced representations and projective modules, Ann. of Math., 71, 1960, 552-578).

If R is a commutative ring, we have the subcategories $\mathcal{S}_R$, $\mathcal{C}_R$ of $\mathfrak{M}_R$ consisting of faithfully projective resp. invertible R-modules: $\otimes_R$ gives a product for each of these. The relationships between the K_i of $\mathfrak{M}_R$, $\mathcal{S}_R$ and $\mathcal{C}_R$ are fully discussed in [B]; when G acts as group of automorphisms of R, it acts on all these. We also have the equivariant groups: $K_0(\text{Rep}\ \mathscr{C}_R)$ will be computed in our next paper. Further examples arise from categories of central separable R-algebras: if morphisms are isomorphisms over R (twisted by g ∈ G for morphisms of grade g), we have a category $\mathscr{A}_R$ and a functor $\text{End}_R : \mathscr{S}_R \to \mathscr{A}_R$ which is naturally cofinal. We do not discuss these examples here, but mention them to show that G-graded categories with product do arise in a number of interesting situations.

King's College,
London.
Liverpool University.

TRIANGULATED CATEGORIES AND ALGEBRAIC K-THEORY

by I. Bucur

The purpose of this lecture is to expose some results of Grothendieck and Verdier concerning the possibility of using the notions of triangulated category and derived category in Algebraic K-Theory.

In Algebraic Topology, the most important topological invariants of a topological space X are the homology and cohomology groups $H^i(X, \mathcal{F})$, with coefficients in a sheaf $\mathcal{F}$ of abelian groups. In order to obtain these groups one usually begins by defining a complex of abelian groups:

$$\cdots \to K_i \xrightarrow{d_i^K} K_{i+1} \to \cdots$$

whose homology groups are precisely $H^i(X, \mathcal{F})$. The idea of Grothendieck and Verdier was to consider, for a topological space, not only the groups $H^i(X)$ and $H_i(X)$ but - for a precise homology theory - one entire complex of abelian groups or better of A-modules whose homology groups are $H^i(X)$ or $H_i(X)$.

Generally, for a given homology theory, these complexes are not uniquely determined, but usually they are quasi-isomorphic. Precisely if X and Y are two complexes of A-modules, we say that they are quasi-isomorphic if there exists a morphism $u : X \to Y$ of complexes which induces an isomorphism on cohomology. In this way one arrives at the derived category, D(A), of an abelian category A.

The derived category D(A) is obtained as follows: one first considers the category K(A), whose objects are complexes of elements of A, and whose morphisms are homotopy equivalence classes of morphisms of complexes. The category D(A) is a category of fractions, obtained from K(A) by inversing formally every morphism in K(A) which is a quasi-isomorphism, i.e. which induces an isomorphism of cohomology. Consequently the categories K(A) and D(A) have the same objects but each quasi-isomorphism in K(A) becomes an isomorphism in D(A).

The category D(A) has a certain extra structure - the structure of a triangulated category - which allows the extension to complexes of the usual formalism of exact sequences.

We shall not give the actual definition of a triangulated category* but we shall say only the following:

$\mathcal{A}_0$ is a pre-additive and graded category with a translation functor $T : \mathcal{A}_0 \to \mathcal{A}_0$, $(T(X) = X[1],\ T(u) = u[1])$. A <u>triangle</u> of $\mathcal{A}_0$ is a collection (X, Y, Z, u, v, w), where X, Y, Z are objects of $\mathcal{A}_0$, and u, v, w are morphisms as follows: $u : X \to Y$, $v : Y \to Z$, $w : Z \to T(X)$. A triangle is usually written

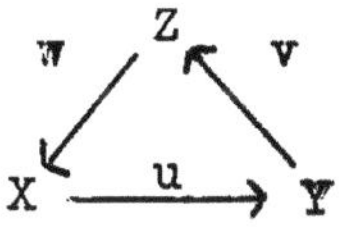

The morphism u is called the base of the triangle.

A <u>morphism</u> of triangles

$(X, Y, Z, u, v, w) \to (X', Y', Z', u', v', w')$ is a commutative diagram

---

* It is too technical. One <u>supposes</u> known the theory of 2-categories and 2-functors. [cf. [2]]

$$\begin{array}{ccccccc}
X & \xrightarrow{u} & Y & \xrightarrow{v} & Z & \xrightarrow{w} & X[1] \\
{\scriptstyle f}\downarrow & & {\scriptstyle g}\downarrow & & {\scriptstyle h}\downarrow & & \downarrow{\scriptstyle f[1]} \\
X' & \xrightarrow{u'} & Y' & \xrightarrow{v'} & Z' & \xrightarrow{w'} & X'[1]
\end{array}$$

One obtains in this way an additive, graded category $\mathcal{T}(\mathcal{A}_0)$, the category of triangles of $\mathcal{A}_0$.

A <u>triangulated category</u> is a graded, additive functor

$$\partial : \mathcal{A}_1 \to \mathcal{T}(\mathcal{A}_0)$$

from one pre-additive category $\mathcal{A}_1$ to the category of triangles of a pre-additive category $\mathcal{A}_0$ so that the following axioms are satisfied:

T1. For each $A \in Ob(\mathcal{A}_0)$ there exists $\Delta \in Ob(\mathcal{A}_1)$ such that $\partial(\Delta)$ is isomorphic with the triangle $(A, A, 0, id_A, 0, 0)$:

$$\begin{array}{ccc}
 & 0 & \\
\swarrow & & \nwarrow \\
A & \xrightarrow{id_A} & A
\end{array}$$

T2. Given two objects Δ_0 and Δ_1 of $\mathcal{A}_1$ such that $\partial(\Delta_0) = (A, A, 0, id_A, 0, 0)$, $\partial(\Delta_1) = (X, Y, Z, u, v, w)$, the function

$$Hom_{\mathcal{A}_1}(\Delta_0, \Delta_1) \to Hom_{\mathcal{A}_0}(A, X)$$

is an isomorphism.

T3. For each morphism u of $\mathcal{A}_0$ there exists $\Delta \in Ob(\mathcal{A}_1)$ such that u is isomorphic with the base of the triangle $\partial(\Delta)$.

T4. Let Δ_1 and Δ_2 be two objects of $\mathcal{A}_1$ and u_1, u_2 the bases of the triangles $\partial(\Delta_1)$ and $\partial(\Delta_2)$. Then <u>every</u> commutative diagram

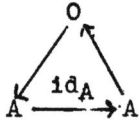

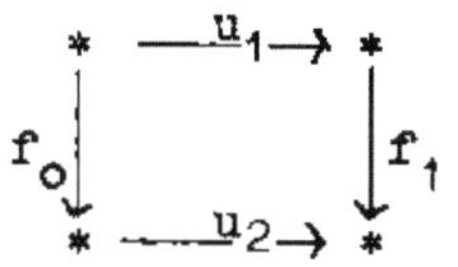

is induced by a morphism f from Δ_1 to Δ_2.

T5. If $(X, Y, Z, u, v, w) \in \mathcal{T}(\mathcal{A}_0)$ is isomorphic with a triangle of the form $\partial(\Delta)$, then the triangle $(Y, Z, X[1], v, w, -u[1])$ is also isomorphic with a triangle of the form $\partial(\Delta')$.

DEFINITION. The triangles of $\mathcal{A}_0$ which belong to the essential image of ∂ will be called <u>distinguished</u> triangles. If $u : \Delta \to \Delta'$ is a morphism of the category $\mathcal{A}_1$, then the components of the morphism $\partial(u)$ will be denoted by $u_0, u_1, u_2, \ldots$

1. Additive functions and theories of determinants.

DEFINITION 1.1. Let $\mathcal{A}_* = (\mathcal{A}_0, \mathcal{A}_1, \partial)$ be a triangulated category. A function $f : \mathrm{Ob}(\mathcal{A}_0) \to G$ from the set of objects of $\mathcal{A}_0$ to the abelian group G is called an <u>additive</u> function on $\mathcal{A}_*$ with values in G if for every distinguished triangle

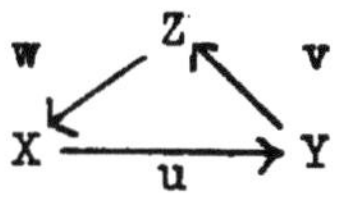

the following relation is true

$$f(X) + f(Z) = f(Y).$$

The set of all additive functions on $\mathcal{A}_*$ with values in G is naturally an abelian group so that we obtain a functor $\mathrm{add}(\mathcal{A}_*) : Ab \to Ab$.

This functor is representable by a pair $(k^o(\mathcal{A}_*), cl)$ where $k^o(\mathcal{A}_*)$ is an abelian group called the Grothendieck group of $\mathcal{A}_*$ and cl is a universally additive function on $\mathcal{A}_*$ with values in $k^o(\mathcal{A}_*)$.

One can also define the second K-group of the triangulated category $\mathcal{A}_*$.

DEFINITION 1.2. We call a function $f : Ob(\underline{Aut}(\mathcal{A}_o)) \to G$ a theory of determinants on $\mathcal{A}_*$ with values in the abelian group G if the following axioms are satisfied.

1. $f(uv) = f(u) + f(v)$
2. If $u : \Delta \to \Delta$ is an automorphism of a triangle, then

$$f(u_1) = f(u_o) + f(u_2).$$

The standard proof shows that there exists a universal theory of determinants

$$D : Ob(\underline{Aut}(\mathcal{A}_o)) \to k^1(\mathcal{A}_*)$$

and the abelian group $k^1(\mathcal{A}_*)$ will be called the Whitehead group of $\mathcal{A}_*$.

Using the notion of a multiplicative category it is possible to obtain at the same time the groups of type k^o and k^1.

DEFINITION 1.3. A <u>multiplicative category</u> ([1]) is given by a system $\mathfrak{M} = (\mathscr{C}, \otimes, \Lambda, \theta, \delta)$, where

1. $\mathscr{C}$ is a category

2. $\otimes : \mathscr{C} \times \mathscr{C} \to \mathscr{C}$ is a functor

3. Λ is an object of $\mathscr{C}$, called the unit (for $\otimes$)

4. θ is a natural isomorphism between the functors:

$\otimes \circ (1_{\mathscr{C}} \times \otimes)$, $\otimes \circ (\otimes \times 1_{\mathscr{C}}) : \mathscr{C} \times \mathscr{C} \times \mathscr{C} \longrightarrow \mathscr{C}$

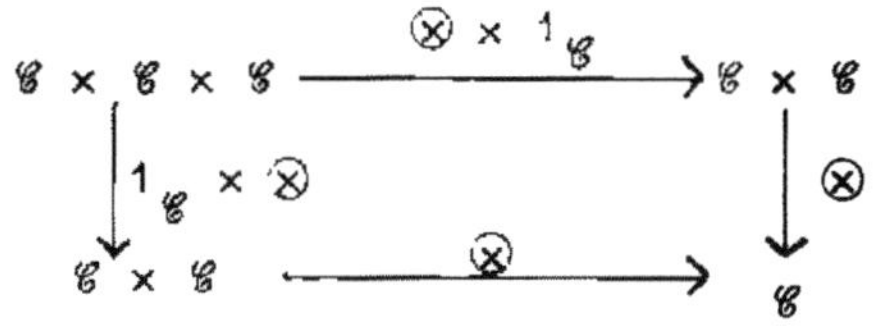

5'. γ is natural isomorphism between the functors:

$1_{\mathscr{C}}$, $\otimes \circ (\Lambda, 1_{\mathscr{C}}) : \mathscr{C} \longrightarrow \mathscr{C}$

($(\Lambda, 1_{\mathscr{C}}) : \mathscr{C} \longrightarrow \mathscr{C} \times \mathscr{C}$ is defined by the formulae:

$(\Lambda, 1_{\mathscr{C}})(A) = \Lambda \otimes A$

$(\Lambda, 1_{\mathscr{C}})(u) = 1 \otimes u$.)

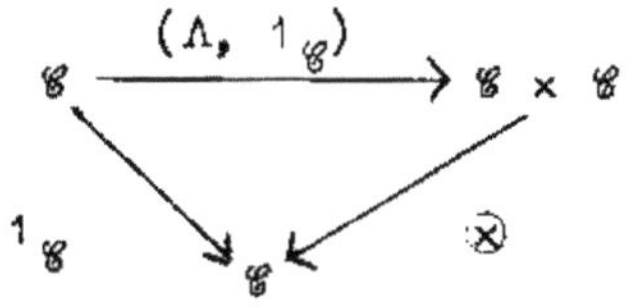

5". δ is a natural isomorphism between the functors

$1_{\mathscr{C}}$, $\otimes \circ (1_{\mathscr{C}}, \Lambda) : \mathscr{C} \to \mathscr{C}$

The notations $\otimes$, Λ remind us of the principal example which we have in mind.

The multiplicative category $\mathfrak{M}$ is called <u>commutative</u> if there exists a natural isomorphism $\sigma : \otimes \to \otimes \circ S$ where $S : \mathscr{C} \times \mathscr{C} \to \mathscr{C} \times \mathscr{C}$, $S(A_1, A_2) = (A_2, A_1)$, $S(u_1, u_2) = (u_2, u_1)$ is the permutation functor.

DEFINITION 1.4. A multiplicative category $\mathfrak{M} = (\mathscr{C}, \otimes, \Lambda, \gamma, \delta)$ is a <u>Picard category</u> if it is commutative and the following conditions are fulfilled:

i) The morphisms of $\mathscr{C}$ are isomorphisms.

ii) The each object B of $\mathscr{C}$ the functor

$$\mathscr{C}^o \longrightarrow \mathrm{Ens}$$

$$A \longmapsto \mathrm{Hom}_{\mathscr{C}}(A \otimes B, \Lambda)$$

is representable by a pair $(\underline{\mathrm{Hom}}_{\mathscr{C}}(B, \Lambda)\ \nu_B)$ where $\underline{\mathrm{Hom}}_{\mathscr{C}}(B, \Lambda)$ is an object of $\mathscr{C}$ and $\nu_B : \underline{\mathrm{Hom}}_{\mathscr{C}}(B, \Lambda) \otimes B \to \Lambda$ is a morphism of $\mathscr{C}$, consequently an isomorphism. We have a natural isomorphism $\mathrm{Hom}_{\mathscr{C}}(A \otimes B, \Lambda) \simeq \mathrm{Hom}_{\mathscr{C}}(A, \underline{\mathrm{Hom}}_{\mathscr{C}}(B, \Lambda))$. A standard example of a Picard category is $\underline{\mathrm{Pic}}(\Lambda)$ where Λ is a commutative ring. This category is generated by the projective Λ-modules of rank one. Let $\pi_o(\mathfrak{M})$ be the set of classes of isomorphic objects of $\mathscr{C}$. This set has the natural structure of an abelian group induced by the multiplication of $\mathfrak{M}$.

PROPOSITION 1.5. <u>If</u> $\mathfrak{M} = (\mathscr{C}, \otimes, \Lambda, \gamma, \delta)$ <u>is a Picard category and</u> $P \in \mathrm{Ob}(\mathscr{C})$ <u>then</u> $\pi_1(P) = \mathrm{Aut}(P) = \mathrm{Hom}_{\mathscr{C}}(P, P)$ <u>is an abelian group and naturally isomorphic with</u>

$\pi_1(\Lambda) = \mathrm{Aut}(\Lambda, \Lambda) = \pi_1(\mathfrak{M})$.

PROOF. The isomorphism between the group $\pi_1(P)$ and $\pi_1(\Lambda)$ is the following:

$$\begin{array}{ccc} \pi_1(P) = \mathrm{Hom}_{\mathscr{C}}(P, P) & \longrightarrow & \mathrm{Hom}_{\mathscr{C}}(\underline{\mathrm{Hom}}_{\mathscr{C}}(P, \Lambda) \otimes P, \underline{\mathrm{Hom}}_{\mathscr{C}}(P, \Lambda) \otimes P) \\ u & \longmapsto & 1 \otimes u \\ & & \downarrow \\ & & \mathrm{Hom}_{\mathscr{C}}(\Lambda, \Lambda) = \pi_1(\Lambda) \end{array}$$

We often denote the object $\underline{\mathrm{Hom}}_{\mathcal{C}}(P, \Lambda)$ by P^{-1}. It is clear that $P \longmapsto P^{-1}$ is a functor. The homomorphism described is indeed an isomorphism because there exists its inverse:

$$\mathrm{Hom}_{\mathcal{C}}(\Lambda, \Lambda) \longrightarrow \mathrm{Hom}_{\mathcal{C}}(P \otimes \underline{\mathrm{Hom}}_{\mathcal{C}}(P, \Lambda), \Lambda) \longrightarrow$$

$$\mathrm{Hom}_{\mathcal{C}}(P \otimes \underline{\mathrm{Hom}}_{\mathcal{C}}(P, \Lambda) \otimes P, \Lambda \otimes P) \simeq \mathrm{Hom}_{\mathcal{C}}(P, P)$$

The second assertion of the proposition is proved.

For the first part we shall show that $\pi_1(\Lambda)$ is a group in the category of groups, and then that it is an abelian group. It is enough to define the mapping:

$$\mu : \pi_1(\Lambda) \times \pi_1(\Lambda) \longrightarrow \pi_1(\Lambda)$$

by the formula:

$$u \top v = \mu(u, v) = \gamma(\Lambda)(u \otimes v)(\gamma(\Lambda))^{-1}:$$

$$\begin{array}{ccc} \Lambda \otimes \Lambda & \xrightarrow{\ u \otimes v\ } & \Lambda \otimes \Lambda \\ {\scriptstyle \gamma(\Lambda)} \downarrow & & \downarrow {\scriptstyle \gamma(\Lambda)} \\ \Lambda & \xrightarrow{\ u \top v\ } & \Lambda \end{array}$$

We shall verify that μ is a morphism in the category of groups:

$$\mu((u, v).(u', v')) = \mu(u, v).\mu(u', v').$$

Indeed

$$\mu((u, v).(u', v')) = \mu(uu', vv') = \gamma(\Lambda)(uu' \otimes vv')(\gamma(\Lambda))^{-1}$$

$$\begin{aligned} \mu(u, v).\mu(u', v') &= \gamma(\Lambda)(u \otimes v)(\gamma(\Lambda))^{-1}(\gamma(\Lambda))(u' \otimes v')(\gamma(\Lambda))^{-1} \\ &= \gamma(\Lambda)(u \otimes v)(u' \otimes v')(\gamma(\Lambda))^{-1} \\ &= \gamma(\Lambda)(uu' \otimes vv')(\gamma(\Lambda))^{-1}. \end{aligned}$$

DEFINITION 1.6. Let $\mathcal{A}_* = (\mathcal{A}_0, \mathcal{A}_1, \partial)$ be a triangulated category and $\mathcal{P} = (\mathcal{C}, \otimes, \Lambda, \gamma, \delta)$ a Picard category. A general theory of determinants on $\mathcal{A}_*$ with values in $\mathcal{P}$ is a pair (D, c) where

$$D : \mathcal{A}_{o,is} \longrightarrow \mathcal{C}$$

is a functor and

$$c : Ob(\mathcal{A}_1) \longrightarrow \text{the class of morphisms of } \mathcal{C}$$

such that the following conditions are fulfilled:

(i) $D(0) = \Lambda$ ($\mathcal{A}_o$ is a pre-additive category)

(ii) if $\Delta \in Ob(\mathcal{A}_1)$ and $\partial(\Delta) = (A, B, C, u, v, w)$

$c(\Delta) : D(A) \otimes D(C) \xrightarrow{\sim} D(B)$

if $\partial(\Delta) = (A, A, 0, id_A, 0, 0)$ then $c(\Delta) = \delta(D(A))$

(iii) if $f : \Delta \to \Delta'$ is an isomorphism of triangles,

$\partial(\Delta') = (A', B', C', u', v', w')$, $\partial f = (f_o, f_1, f_2)$

then the following diagram is commutative.

$$\begin{array}{ccc} D(A) \otimes D(C) & \xrightarrow{c(\Delta)} & D(B) \\ {\scriptstyle D(f_o) \otimes D(f_2)} \downarrow & & \downarrow {\scriptstyle D(f_1)} \\ D(A') \otimes D(C') & \xrightarrow[c(\Delta')]{} & D(B') \end{array}$$

PROPOSITION 1.7. Every general theory of determinants $D : \mathcal{A}_{o,is} \to \mathcal{C}$ defined on $\mathcal{A}_* = (\mathcal{A}_i)_{i \geq 0}$ with values in $\mathcal{P} = (\mathcal{C}, \otimes, \Lambda, \gamma, \delta)$ generates an additive function $f_D : Ob(\mathcal{A}_o) \to \pi_o(\mathcal{P})$

$$f_D(X) = \text{the class of } D(X) \text{ in } \pi_o(\mathcal{P})$$

and a theory of determinants $\delta_D : Ob(\underline{Aut}(\mathcal{A}_o)) \to \pi_1(\mathcal{P})$ where

$\delta_D(P \overset{u}{\to} P)$ = the image of u in $\pi_1(\mathcal{P})$ by the canonical homomorphism $Aut(\mathcal{P}) \to Aut(\Lambda) = \pi_1(\mathcal{P})$.

PROOF. Let $\Delta \in Ob(\mathcal{A}_1)$ be a triangle so that $\partial(\Delta) = (X, Y, Z, u, v, w)$. Then there exists (Definition 1.6.(ii)) an isomorphism

$$c(\Delta) : D(X) \otimes D(Z) \xrightarrow{\sim} D(Y)$$

and consequently we have

$$f_D(X) + f_D(Z) = f_D(Y).$$

If $f : \Delta \to \Delta$ is an automorphism then condition (iii) implies

$$\delta_D(f_o) + \delta_D(f_2) = \delta_D(f_1).$$

It is clear that if $u, v : P \to P$ are automorphisms of P we have

$$\delta_D(uv) = \delta_D(u) + \delta_D(v).$$

PROBLEM (Grothendieck). Let $\mathcal{A} = (\mathcal{A}_n)_{n\geq o}$ be a triangulated category. Does there exist a Picard category $\mathcal{P} = (\mathcal{C}, \otimes, \Lambda, \gamma, \delta)$ and a general theory of determinants

$$D : \mathcal{A}_{o,is} \to \mathcal{C}$$

such that the associated additive function

$$f_D : Ob(\mathcal{A}_o) \to \pi_o(\mathcal{P})$$

and the associated theory of determinants

$$\delta_D : Ob(\underline{Aut}(\mathcal{A}_o)) \to \pi_1(\mathcal{P})$$

are universal?

2. Comparison with classical K-theory.

2.1. Let C_o be a full sub-category of an abelian category C. One can define the abelian group $A_{C_o}(G)$ of additive functions on C_o with values in the abelian group G and, of course, the group $D_{C_o}(G)$ of theories of determinants on $\underline{Aut}(C_o)$ with values in G. These functors

$$D_{C_o}, A_{C_o} : Ab \to Ab$$

allow us to define the groups K(D) or $K_o(D)$ and W(D) or $K_1(D)$. Assume now that the sub-category C_o of C satisfies the following conditions:

(A) C_o is stable for finite direct sums.

(B) If X, Y $\in Ob(C_o)$ and $X \overset{u}{\to} Y$ is an epimorphism of C, then $Ker(u) \in Ob(C_o)$

(C) Each diagram

$$\begin{array}{ccc} A & \xrightarrow{u} & B \\ & & \uparrow \\ & & L \end{array}$$

where u is an epimorphism and $L \in Ob(C_o)$, can be embedded in a commutative diagram

$$\begin{array}{ccc} A & \xrightarrow{u} & B \\ \uparrow & & \uparrow \\ L_1 & \xrightarrow{w} & L \end{array}$$

such that $L_1 \in Ob(C_o)$ and w is an epimorphism.

In this case it is possible to define the triangulated categories $\mathfrak{D}_o^-(C)_{coh}$, $\mathfrak{D}_o(C)_{parf}$ as sub-categories of the derived category $\mathfrak{D}(C)$ of the abelian category C. First the derived category $\mathfrak{D}(C) = (\mathscr{C}_o, \mathscr{C}_1, \partial_1)$ is defined as follows: $\mathscr{C}_o = D(C)$ is the additive category obtained by localisation from K(C) - the category of complexes with morphisms the classes of homotopy - by the multiplicative system of quasi-isomorphisms. Consequently a morphism from X to Y in D(C) is represented by a diagram

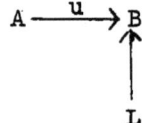

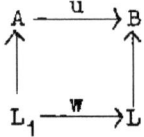

$$X \xleftarrow{\;s\;} X' \xrightarrow{\;a\;} Y$$

with s a quasi-isomorphism. This diagram defines the same morphism as another one

$$X \xleftarrow{\;t\;} X'' \xrightarrow{\;b\;} Y$$

if and only if there is a quasi-isomorphism $u : X''' \to X$ and morphisms $f : X''' \to X'$, $g : X''' \to X''$ such that the following diagrams are commutative:

$$\begin{array}{ccc} X''' & \xrightarrow{f} & X' \\ & {\scriptstyle u}\searrow \quad \swarrow{\scriptstyle s} & \\ & X & \end{array} \qquad \begin{array}{ccc} X''' & \xrightarrow{g} & X'' \\ & {\scriptstyle u}\searrow \quad \swarrow{\scriptstyle t} & \\ & X & \end{array} \qquad \begin{array}{ccc} X''' & \xrightarrow{f} & X' \\ {\scriptstyle g}\downarrow & & \downarrow{\scriptstyle a} \\ X'' & \xrightarrow{b} & Y \end{array}$$

The objects of the category $\mathscr{C}_1$ are the pairs $(X, Y, \varphi : Y \to X[1])$ where φ is a morphism of complexes. If $\Delta = (X, Y, \varphi : Y \to X[1])$, $\Delta' = (U, V, \psi : V \to U[1])$ are two objects of $\mathscr{C}_1$, a morphism $f : \Delta \to \Delta'$ is given by a pair (u, v) in the category D(C), where $u : X \to U$ and $v : Y \to V$ are such that the following diagram commutes in D(C).

$$* \qquad \begin{array}{ccc} Y & \xrightarrow{\varphi} & X[1] \\ {\scriptstyle v}\downarrow & & \downarrow{\scriptstyle u[1]} \\ V & \xrightarrow{\psi} & Y[1] \end{array}$$

The functor $\partial_1 : \mathscr{C}_1 \to D(C)$ is defined as follows.

$\partial_1(X, Y, \varphi : Y \to X[1]) = (X, X \oplus_\varphi Y, Y, i_X, p_Y, \varphi)$, where $(X \oplus_\varphi Y)_i = X_i \oplus Y_i$

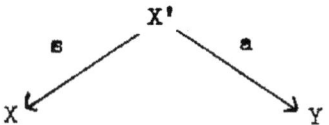

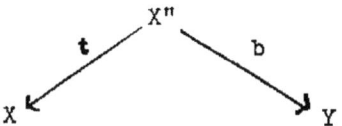

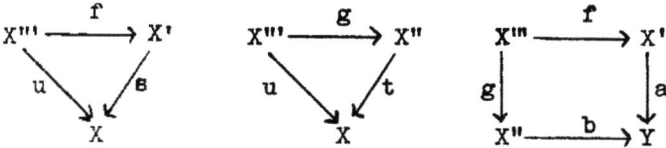

$\partial_i : X_i \oplus Y_i \to X_{i+1} \oplus Y_{i+1}$ has the following associated matrix:

$$\begin{pmatrix} d_i^X & \varphi_i \\ 0 & d_i^Y \end{pmatrix}$$

The components f_o, f_1, f_2 of $\partial_1(f)$ for $f = (u, v)$ are the following:

$f_o = u$

$f_2 = v$

f_1 = the extension to $X \oplus_\varphi Y$, $U \oplus_\psi V$ of u and v (in D(C)).

It is well to remark that there exist morphisms between triangles in D(C) which are not induced by morphisms of objects of $\mathscr{C}_1$.

If we consider in the above construction only the complexes bounded below, we obtain the categories $\mathfrak{D}^+(C)$, $D^+(C)$. Similarly we define the categories $\mathfrak{D}^-(C)$, $\mathfrak{D}^b(C)$, $D^-(C)$, $D^b(C)$ by taking complexes bounded above, or bounded on both sides, respectively.

More generally, by taking complexes whose components are objects of C_o we define the category $D_o^*(C)$ $(* = +, -, \varphi)$ and a functor

$$i^* : D_o^*(C) \to D^*(C).$$

Under our assumptions (A) (B), (C), this functor is a full embedding in the cases $* = -$ or b.

DEFINITION 2.1.1. (i) The essential image of $D_o^-(C)$ by the functor i^- will be denoted by $D_o^-(C)_{coh}$ and its objects will be called C_o-pseudo-coherent complexes of C.

(ii) The essential image of $D^b_o(C)$ by the functor i^b will be denoted by $D_o(C)_{parf}$ and its objects will be called C_o-perfect complexes of C.

Obviously every C_o-pseudo-coherent complex is a C_o-perfect complex.

EXAMPLE 2.1.2. Let C = mod(A) (left A-modules for instance) and C_o = P(A) the full subcategory generated by the projective A-modules of finite type. This subcategory satisfies conditions (A), (B) and (C). In this case we shall use the following notations:

$$D^-(A)_{coh} = D^-_o(C)_{coh}.$$

$$\mathrm{Parf}(A) = D(A)_{parf} = D_o(C)_{parf}.$$

We shall drop the notation C_o in the case of the category mod(A) when C_o = P(A).

PROPOSITION 2.1.3. *If A is a ring and L a complex of A-modules, the following conditions are equivalent.*

(a) *L is a perfect complex of A-modules.*

(b) *L is pseudo-coherent with bounded cohomology and of finite Tor-dimension.*

PROPOSITION 2.1.4. *There exist two triangulated categories* $\mathcal{D}^-_o(C)_{coh} = (\mathcal{A}_o, \mathcal{A}_1, \delta_1)$, $\mathcal{D}_o(C)_{parf} = (\mathcal{B}_o, \mathcal{B}_1, \delta'_1)$ *such that* $\mathcal{A}_o = D^-_o(C)_{coh}$, $\mathcal{B}_o = D_o(C)_{parf}$.

PROOF. If $\mathcal{D}(C) = (\mathcal{C}_o, \mathcal{C}_1\ \ \partial_1)$ is the derived category of C, it is enough to take $\mathcal{A}_1$ (resp. $\mathcal{B}_1$) the sub-category of $\mathcal{C}_1$ generated by the objects Δ of $\mathcal{C}_1$ such that $\partial_1(\Delta) \in Ob(D_o^-(C)_{coh})$ (resp. $\partial_1(\Delta) \in Ob(D_o(C)_{parf})$. In the case of $C = mod(A)$, $C_o = P(A)$ we shall denote $\mathcal{P}arf(A) = \mathcal{D}_o(C)_{parf}$.

2.2. Let G be an abelian group and

$$f : Ob(D_o(C)_{parf}) \to G$$

an additive function on the triangulated category $\mathcal{D}_o(C)_{parf}$ with values in G. For each object X of C we denote by $X^\bullet$ the complex of the category C having all components zero except in dimension zero which is X.

$$X^\bullet = (\ldots \to 0 \to \ldots \to 0 \to X \to 0 \to \ldots)$$

If we define $\overline{f} : Ob(C_o) \longrightarrow G$ by

$$\overline{f}(X) = f(X^\bullet)$$

we obviously obtain an additive function defined on the category C_o with values in the group G and consequently a homomorphism of groups:

$$\mu : K(C_o) \longrightarrow K(\mathcal{D}_o(C)_{parf})$$

PROPOSITION 2.2.1. <u>The homomorphism</u> μ <u>is in fact an isomorphism. In particular for every ring</u> A <u>there exists a canonical isomorphism</u>:

$$K(A) \xrightarrow{\sim} K(\mathcal{P}arf(A)).$$

Proposition 2.2.1 shows that it is important to have criteria allowing us to obtain perfect complexes. Proposition 2.1.3

is a first result on this way.

As a first example it is easy to see that the results of Verdier over constructible sheaves (Seminaire Bourbaki, Nov. 1965, Exposé 300) combined with proposition 2.2.1 allow us to obtain the following partial answer:

PROPOSITION 2.2.2. Let X be a compact space of finite dimension, G a finite group which operates on X. A a noetherian commutative ring and $\mathcal{F}$ a cohomologically constructible sheaf of A-modules on X. Assume also that G operates on $\mathcal{F}$ compatibly with its operation on X. Consequently the full cohomological complex $R\Gamma_X(\mathcal{F})$ is a complex of $A[G]$-modules. For every $x \in X$, denote by G_x the stability group of x, then the stalk $\mathcal{F}_x$ of $\mathcal{F}$ over the point x is a $A[G_x]$-module. If we suppose that $\mathcal{F}_x$ is a flat $A[G_x]$-module, then the complex $R\Gamma_X(\mathcal{F})$ is a perfect complex of $A[G]$-modules.

PROOF. (a) We want to show first that $R\Gamma_X(\mathcal{F})$ is a finite Tor-dimensional complex of $A[G]$-modules. Indeed, if $Y = X/G$ is the space of orbits and $p : X \to Y$ the natural map assigning to $x \in X$ its orbit, we have the following isomorphism of complexes of $A[G]$-modules:

$$R\Gamma_X(\mathcal{F}) \simeq R\Gamma_Y(Rp_*(\mathcal{F}))$$

But $Rp_*(\mathcal{F})$ is isomorphic with $p_*(\mathcal{F})$. This last sheaf is a sheaf of $A[G]$-modules on the space Y and our assumptions imply that its stalks are flat $A[G]$-modules. Therefore the complex $Rp_*(\mathcal{F})$ of $A[G]$-modules is of finite Tor-dimension. This implies by the simple application of the "Projection formula" that $R\Gamma_Y(Rp_*(\mathcal{F}))$ is of finite Tor-dimension.

(b) The complex $R\Gamma_X(\mathcal{F})$ is also pseudo-coherent as a complex of A[G]-modules. This results from the hypotheses of constructibility.

Indeed the result of Verdier (in fact the general form of the theorem of Willder) implies that the cohomology groups of $R\Gamma_X(\mathcal{F})$ are of finite type. But this is enough because A[G] is noetherien. The determination of the class $cl_{K(A[G])}(R\Gamma_X(\mathcal{F}))$ in terms of the classical invariants of X, G and F is a very interesting but more difficult problem except in the case when the set of fixed-points of G is isolated. Another example is connected with the Wall obstruction to homotopical finiteness of a CW-complex.

<u>DEFINITION 2.2.3</u>. Let X be a connected CW-complex, $\tilde{X}$ its universal covering space and $\pi = \pi_1(X, x_0)$ the fundamental group of X at the point x_0. We define the complex

$$w(X) = (C^n(\tilde{X}), d_n)$$

of $Z[\pi]$-modules in the following way:

First put $C^n(\tilde{X}) = H_n(\tilde{X}^n, \tilde{X}^{n-1}; Z)$, where $\tilde{X}^n$ is the n-dimensional skeleton of $\tilde{X}$. Of course $C^n(\tilde{X})$ is considered as a π-module. The homomorphism $d_n : C^n(\tilde{X}) \to C^{n-1}(\tilde{X})$ is that associated with the triple $(\tilde{X}^n, \tilde{X}^{n-1}, \tilde{X}^{n-2})$.

The complex $w(X)$ will be considered as an object of $D^-(Z[\pi])$.

PROPOSITION 2.2.4. <u>Let</u> L <u>be a sub-complex of</u> X <u>such that</u>

the homomorphism $\pi_1(L) \to \pi_1(X)$ *induced by the canonical injection of* L *in* X *is an isomorphism. In this way* $w(L)$ *is also an object of* $D^-(Z[\pi])$ *and the canonical injection induces a morphism* $w(L) \to w(X)$, *in* $D^-(Z[\pi])$ *such that the triangle*

$$\begin{array}{ccc} & w(X, L) & \\ \swarrow & & \nwarrow \\ w(L) & \longrightarrow & w(X) \end{array}$$

is distinguished, and we have the isomorphisms:

$$H_i(w(X, L)) \simeq H_i(\tilde{X}, \tilde{L}, Z).$$

PROOF. It is enough to remark that, $(\tilde{X}, X, p)$ being the universal covering space of X and $\pi_1(L) \to \pi_1(X)$ an isomorphism, then $\tilde{L} = p^{-1}(L)$.

PROPOSITION 2.2.5. *Let* X *be a* CW-*complex dominated by a finite* CW-*complex and assume there exists an integer* $n \geq 1$ *such that*

(a) $H_i(\tilde{X}, Z) = 0$ *for* $i \geq n + 1$.

and (b) $H^{n+1}(X, \mathcal{A}) = 0$ *for all local coefficients.*

Under these conditions, the complex $w(X)$ *is perfect. We denote the class of this complex in the Grothendieck group* $K(Z[\pi])$ *by* $\omega(X)$. *The image* $\bar{\omega}(X)$ *of* $\omega(X)$ *in the stable Grothendieck group associated to the ring* $Z[\pi]$ *is precisely the Wall-obstruction to homotopical finiteness of* X.

PROOF. There exists a finite polyhedron L of dimension

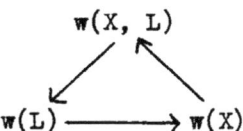

$n - 1$ and a continuous map [8] $\varphi : L \to X$ such that

(i) $\pi_i(\varphi) = 0$ for $i = 1, 2, \ldots, n - 1$.

and (ii) $\pi_n(\varphi)$ is a projective π-module of finite type. (As a matter of fact, the class of this π-module multiplied by $(-1)^n$ is, by definition, the Wall-invariant.)

Obviously we can suppose φ an inclusion. Secondly we can suppose $n \geqslant 2$. For the case $n \leqslant 1$ one argues directly as an Wall's paper. Therefore we have the distinguished triangle:

$$(*) \qquad \begin{array}{ccc} & w(X, L) & \\ {}^{1}\swarrow & & \nwarrow \\ w(L) & \longrightarrow & w(X) \end{array}$$

But the complex L is finite, and therefore $w(L)$ is perfect. Consequently, it is enough to show that $w(X, L)$ is perfect. Using proposition 2.2.4, the Hurewicz isomorphism and condition (ii) one gets:

$$H_i(w(X, L)) = \begin{cases} 0 & \text{if } i \neq n \\ \pi_n(X, L) = \pi_n(\varphi) \simeq H_n(\tilde{X}, \tilde{L}) & \end{cases}$$

For the second part of the proposition we consider once again the distinguished triangle (*) and we remark that the image of the class of $w(L)$ in the stable Grothendieck group is zero because L is a finite complex.

2.3. There exists for theories of determinants a result anologuous to proposition 2.2.1. Obviously we have a canonical homomorphism of groups:

$$(2.3.1) \qquad \sigma : K^1(C_o) \longrightarrow K^1(\mathcal{D}_o(C)_{parf})$$

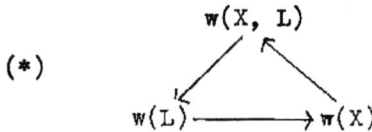

PROPOSITION 2.3.2. (Verdier) *If* $C = \text{mod}(A)$, $C_o = P(A)$, *then the homomorphism* $\sigma(2.3.1)$ *is an isomorphism.*

The proof of this theorem will be a consequence of the following series of lemmas:

LEMMA 2.3.3. *Let* C *be an abelian category and*

$$X = (\dots \to X_n \xrightarrow{d_n^X} X_{n+1} \to \dots \to X_{i-1} \xrightarrow{d_{i-1}^X} X_i \to 0 \to 0 \to \dots)$$

a complex in C *bounded at the right. Assume that* $\varphi_i : P_i \to X_i$ *is an epimorphism, and let* $d_{i-2}^P : X_{i-2} \to P \Pi_{X_i} X_{i-1}$ *be the morphism determined by the conditions:*

$$p_{X_i} d_{i-2}^P = d_{i-2}^X$$

$$p_{P_i} d_{i-2}^P = 0$$

Denote by P *the complex:*

$$P = (\dots \to X_n \xrightarrow{d_n^X} X_{n+1} \to \dots \to X_{i-2} \xrightarrow{d_{i-2}^P} P_i \Pi_{X_i} X_{i-1} \xrightarrow{p_{P_i}} P_i \to 0 \to \dots)$$

There exists a quasi-isomorphism $\psi = (\psi_j)_{j \in Z} : P \to X$ *such that:*

$$\psi_j = 1_{X_j} \text{ for } j \leq i-2$$

$$\psi_{i-1} = p_{X_{i-1}}$$

$$\psi_i = \varphi_i$$

$$\psi_j = 0 \text{ if } j \geq i+1.$$

If

$$X' = (\dots \to X'_n \xrightarrow{d_n^{X'}} X'_{n+1} \to \dots \to X'_{i-1} \xrightarrow{d_{i-1}^{X'}} X'_i \to 0 \to 0 \to \dots)$$

is another complex of C *satisfying the same conditions as* X, $\varphi'_i : P'_i \to X'_i$ *an epimorphism,* P' *the complex associated to*

X' as above, $\psi' = (\psi_i') : P' \to X'$ the corresponding quasi-isomorphism and $u : X \to X'$ and $v_i : P_i \to P_i'$ the morphisms making the following diagram commute.

$$\begin{array}{ccc} P_i & \xrightarrow{\varphi_i} & X_i \\ {\scriptstyle v_i}\downarrow & & \downarrow{\scriptstyle u_i} \\ P_i' & \xrightarrow[\varphi_i']{} & X_i' \end{array}$$

Under these conditions there exists a morphism $v = (v_i)_{i \in Z} : P \to P'$ such that the diagram:

$$\begin{array}{ccc} P & \xrightarrow{\psi} & X \\ {\scriptstyle v}\downarrow & & \downarrow{\scriptstyle u} \\ P' & \xrightarrow{\psi'} & X' \end{array}$$

is commutative.

LEMMA 2.3.4. Let

$$K = (\ldots \to K_n \xrightarrow{d_n^K} K_{n+1} \to \ldots \to K_{i-1} \xrightarrow{d_{i-1}^K} K_i \to 0 \to \ldots)$$

$$L = (\ldots \to L_n \xrightarrow{d_n^L} L_{n+1} \to \ldots \to L_{i-1} \xrightarrow{d_{i-1}^L} L_i \to 0 \to \ldots)$$

be two complexes bounded at the right, $u = (u_j)_{j \in Z} : K \to L$ a morphism of complexes, $\varepsilon : K_i \to H^i(K)$, $\eta : L_i \to H^i(L)$, the canonical morphisms and $v_i : K_i \to L_i$ a morphism such that the following diagram is commutative:

$$\begin{array}{ccc} K_i & \longrightarrow & H^i(K) \\ {\scriptstyle v_i}\downarrow & & \downarrow{\scriptstyle H^i(u)} \\ L_i & \longrightarrow & H^i(L) \end{array}$$

Under these conditions there exists a morphism $w = (w_j)_{j \in Z} : K \to L$ such that the following conditions are satisfied

(a) $w_i = v_i$

(b) $w_j = u_j$ if $j \leq i - 2$

(c) w and u are homotopies

LEMMA 2.3.5.(A. Heller). Let C be an abelian category, $X_i \xrightarrow{a_i} Y_i$, $i = 0, 1$ two epimorphisms and $f : Y_0 \to Y_1$ an isomorphism. If we assume that the objects X_i are projective, there exists $g : X_0 \oplus X_1 \tilde{\to} X_1 \oplus X_0$ such that the following diagram commutes.

$$\begin{array}{ccc} X_0 \oplus X_1 & \xrightarrow{g} & X_1 \oplus X_0 \\ {\scriptstyle (a_0,\ 0)}\downarrow & & \downarrow{\scriptstyle (a_1,\ 0)} \\ Y_0 & \xrightarrow{f} & Y_1 \end{array}$$

LEMMA 2.3.6. Let

$$X = (\ldots \to X_n \xrightarrow{d_n^X} X_{n+1} \to \ldots \to X_{i-1} \xrightarrow{d_{i-1}^X} X_i \to 0 \to \ldots)$$

$$X' = (\ldots \to X'_n \xrightarrow{d_n^{X'}} X'_{n+1} \to \ldots \to X'_{i-1} \to X'_i \to 0 \to \ldots)$$

be two complexes of projective A-modules, of finite type and $u = (u_j)_{j \in Z} : X \to X'$ a morphism of complexes such that $H^i(u) : H^i(X) \to H^i(X')$ is an isomorphism. Under these conditions there exists two complexes

$$P = (\ldots \to P_n \xrightarrow{d_n^P} P_{n-1} \to \ldots \to P_{i-1} \xrightarrow{d_{i-1}^P} P_i \to 0 \to \ldots)$$

$$P' = (\ldots \to P'_n \xrightarrow{d_n^{P'}} P'_{n-1} \to \ldots \to P'_{i-1} \xrightarrow{d_{i-1}^{P'}} P'_i \to 0 \to \ldots)$$

of projective A-modules of finite type, two quasi-isomorphisms $\psi : P \to X$, ψ', $P' \to X'$, and a morphism $v : P \to P'$ such that

<u>the following relations are satisfied</u>.

(a) <u>the diagram</u>

$$\begin{array}{ccc} P & \xrightarrow{v} & P' \\ \psi\downarrow & & \downarrow\psi' \\ X & \xrightarrow{u} & X' \end{array}$$

<u>is commutative up to homotopy</u>.

(b) <u>the morphism</u> $v_i : P_i \to P_i'$ <u>is an isomorphism</u>.

(c) $P_j = X_j$, $P_j' = X_j'$ <u>if</u> $j \leqslant i - 2$.

(d) $P_{i-1} = X_{i-1} \oplus X_i'$, $P_i = X_i \oplus X_i'$

$P_{i-1}' = X_{i-1}' \oplus X_i$, $P_i' = X_i' \oplus X_i$

(e) $d_{i-1}^P = d_{i-1}^X \oplus 1_{X_i'}$, $d_{i-1}^{P'} = d_{i-1}^{X'} \oplus 1_{X_i}$.

DEFINITION 2.3.7. If

$$U = (\ldots \to U_n \xrightarrow{d_n^U} U_{n+1} \to \ldots)$$

is a complex (of objects of an abelian category C) we say that the complex V is obtained from U by an <u>elementary modification</u> if V has the form:

$$V = (\ldots \to U_n \xrightarrow{d_n^U} U_{n+1} \to$$

$$\ldots \to U_{j-1} \xrightarrow{i_j d_{j-1}^U} U_j \oplus X_j \xrightarrow{d_j^U \oplus 1_{X_j}} U_{j+1} \oplus X_j \xrightarrow{d_{j+1}^U p_{j+2}}$$

$$U_{j+2} \xrightarrow{d_{j+2}^U} U_{j+3} \to \ldots)$$

Obviously if V is obtained from U by an elementary modification then there exists a quasi-isomorphism $U \to V$:

$$\begin{array}{ccccc}
\cdots \to & U_n & \xrightarrow{d_n^U} & U_{n+1} & \to \\
& \downarrow {\scriptstyle 1_{U_n}} & & \downarrow {\scriptstyle 1_{U_{n+1}}} & \\
\cdots \to & U_n & \longrightarrow & U_{n+1} & \to
\end{array}$$

$$\begin{array}{ccccccccccc}
\cdots \to & U_{j-1} & \xrightarrow{d_{j-1}} & U_j & \longrightarrow & U_{j+1} & \longrightarrow & U_{j+2} & \longrightarrow & U_{j+3} & \to \cdots \\
& \downarrow {\scriptstyle 1_{U_{j-1}}} & & \downarrow {\scriptstyle i_{U_j}} & & \downarrow {\scriptstyle i_{U_{j+1}}} & & \downarrow {\scriptstyle 1_{U_{j+2}}} & & \downarrow {\scriptstyle 1_{U_{j+3}}} & \\
\cdots \to & U_{j-1} & \longrightarrow & U_j \oplus X_j & \longrightarrow & U_{j+1} \oplus X_j & \longrightarrow & U_{j+2} & \longrightarrow & U_{j+3} & \to \cdots
\end{array}$$

DEFINITION 2.3.8. A complex of A-modules will be called <u>strongly perfect</u> if it is bounded on both sides and its components are projective A-modules of finite type.

LEMMA 2.3.9. (the main lemma). *If* X *is a strongly perfect complex of A-modules and* $u : X \to X$ *a quasi-isomorphism there exist a strongly perfect complex* X_o *of A-modules, an isomorphism* $u_o : X_o \to X_o$ *of complexes and a quasi-isomorphism* $\xi : X \to X_o$ *such that the following diagram is homotopy commutative.*

$$\begin{array}{ccc}
X & \xrightarrow{\xi} & X_o \\
{\scriptstyle u}\downarrow & & \downarrow{\scriptstyle u_o} \\
X & \xrightarrow{\xi} & X_o
\end{array}$$

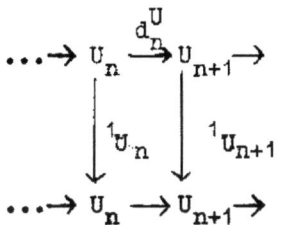

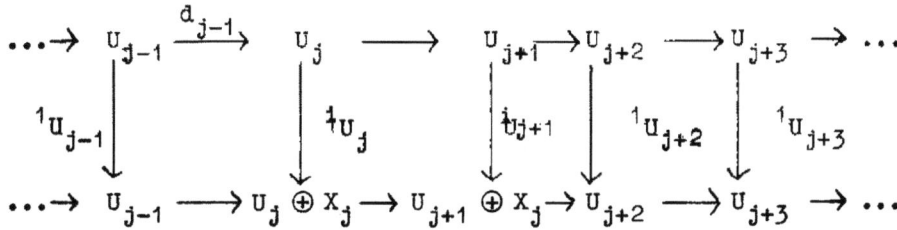

2.4. Applications.

2.4.1. Theory of characteristic polynomials.

Let A be a commutative ring and A[[T]] the ring of formal power series of one variable. We denote by $1 + A[[T]]^+$ the sub-set of A[[T]] of the form $1 + \sum_{i \geq 1} a_i T^i$. The multiplication on A[[T]] induces on $1 + A[[T]]^+$ the structure of an abelian group.

PROPOSITION 2.4.2. <u>There exists a unique function</u>

$$P : Ob(\underline{End}(Parf\ (A))) \to 1 + A[[T]]^+$$

$$u \longmapsto P_u$$

<u>satisfying the following conditions.</u>

(a) <u>If</u> $\mathcal{P}arf(A) = (Parf(A), \mathcal{A}_1, \partial)$, $\Delta \in Ob(A_1)$ <u>and</u> $f : \Delta \to \Delta$ <u>an endomorphism of the triangle</u> Δ, <u>so that the components of</u> $\partial(f)$ <u>are</u> f_0, f_1, f_2, <u>we have</u>

$$P_{f_1} = P_{f_0} . P_{f_2}$$

(b) <u>If</u> $X = (X_i, d_i^X)$ <u>is a strongly perfect complex of</u> A-<u>modules such that</u> $X_i = 0$ <u>for</u> $i \neq 0$ <u>and if</u>

$$u = (u_i)_{i \in Z} : X \to X$$

<u>is an endomorphism of</u> X, <u>then</u> P_u = <u>the usual characteristic polynomial of the endomorphism</u> $u_0 : X_0 \to X_0$.

PROOF. It is easy to see that if $M \in Ob(P(A))$, i.e. is a projective A-module of finite type and $u : M \to M$, then

$$1 - Tu : M \otimes_A A[[T]] \to M \otimes_A A[[T]]$$

is an automorphism of A[[T]]-modules.

Consequently we have the maps:

$$\begin{array}{ccc} Ob(\underline{End}(Parf(A)) & \xrightarrow{\xi} & Ob(\underline{Aut}(Parf(A[[T]])) \\ & & \downarrow \alpha \\ & & K^1(A[[T]]) \end{array}$$

But the notion of determinant induces a homomorphism:

$$\det : K^1(A[[T]]) \longrightarrow U(A[[T]])$$

where $U(A[[T]]$ denotes the group of invertible elements of $A[[T]]$. It is easy to see that the map $(\det)(\alpha\xi)$ factorises through the inclusion $1 + A[[T]]^+ \hookrightarrow U(A[[T]])$ and consequently we get the map

$$P : Ob(\underline{End}(Parf(A))) \longrightarrow 1 + A[[T]]^+.$$

which satisfies the conditions (a) and (b).

REFERENCES.

[1] Bénabou J., Catégories avec multiplication. C.R. Acad. Sci. Paris, 256, 1887-1890, (1963).

[2] Deligne P, Secret Notes on Triangulated and Derived Categories.

[3] Grothendieck A, Séminaire de Géométrie Algebrique, I.H.E.S. - Bures-sur-Yvette, (1965-1966).

[4] Hartshorne R, Residues and Duality, Lecture Notes in Math. 20(1966) Springer.

[7] Verdier J.L, Catégories derivées, Notes polycopiés, I.H.E.S, Bures-sur-Yvette.

[8] Wall C.T.C, Finiteness conditions for CW-complexes, Ann. of Math., 81 (1965) 56-69.

[9] Wall C.T.C, Finiteness conditions for CW-complexes: II, Proc. Roy. Soc., A. 295 (1966) 129-139.

University of Bucharest.

ON MODULES WITH QUADRATIC FORMS

by Anthony Bak

This note communicates a theorem of cancellation (i.e. $M \perp N \cong M' \perp N$ implies $M \cong M'$) for certain quadratic modules over any ring with involution satisfying suitable finiteness conditions, and a theorem classifying the normal subgroups of the unitary groups over such rings.

1. Quadratic Modules.

We fix throughout an algebra A with involution, $a \to \bar{a}$, over a commutative ring k with (compatible) involution; thus $\overline{ab} = \bar{b}\bar{a}$ and $\bar{\bar{a}} = a$ for a, b in A or k. We further fix an element λ in k such that $\lambda\bar{\lambda} = 1$. The involution permits one to convert left modules into right modules and vice-versa, and to speak of conjugate linear homomorphisms of modules. All A-modules considered in this note are to be viewed as right A-modules.

Let F be a left k-module with a conjugate linear automorphism, $B \mapsto \bar{B}$, such that $\bar{\bar{B}} = B$. Then $B \mapsto \lambda\bar{B}$ is also conjugate linear of order two (because $\lambda\bar{\lambda} = 1$) and we shall write $S_\lambda(B) = B + \lambda\bar{B}$. Clearly $S_\lambda(F)$ is a subgroup of $S^\lambda(F) = \{B \in F : B = \lambda\bar{B}\} = \mathrm{Ker}(S_{-\lambda})$. In fact these are modules over the fixed ring $k_1 = S^1(k)$ of the involution.

Now let F = the left k-module of sesquilinear forms B on a right A-module M. Recall that B is a biadditive map $M \times M \to A$ such that $B(ma, nb) = \bar{a}B(m, n)b$ for all $m, n \in M$, $a, b \in A$. We define $\bar{B}$ by $\bar{B}(m, n) = \overline{B(n, m)}$. B is said to be non-singular if the two homomorphisms $M \to M^* = \mathrm{Hom}_A(M, A)$ defined by $m \longmapsto B(m, \ )$ and $m \longmapsto \overline{B}(m, \ )$ are isomorphisms. A <u>quadratic form</u> on M is an element of $F/S_{-\lambda}(F)$. If $B \in F$ we denote the corresponding quadratic form by [B]. In view of the fact $S_\lambda . S_{-\lambda} = 0$ the map S_λ induces a homomorphism $F/S_{-\lambda}(F) \to S^\lambda(F)$ with kernel $S^{-\lambda}(F)/S_{-\lambda}(F)$. We call $B + \lambda\overline{B}$ the <u>λ-hermitian form</u> associated to [B] and we say [B] is non-singular if the form $B + \lambda\overline{B}$ is non-singular. We often write $\langle m, n\rangle_B = B(m, n) + \lambda\overline{B(n, m)}$.

A <u>quadratic module</u> is a pair (M, [B]) consisting of a right A-module M and a quadratic form [B] on M. The (orthogonal) sum $(M, [B]) \perp (N, [C])$ of two quadratic modules is the quadratic module $(M \oplus N, [B \oplus C])$; it is clearly well defined. The quadratic module (M, [B]) is called <u>non-singular</u> if M is finitely generated projective and $\langle , \rangle_B$ is non-singular. (N, [C]) is called a <u>subspace</u> of (M, [B]) iff N is a submodule of M and $[C] = [B|_N]$.

We make the quadratic modules the objects of a category as follows. To define morphisms we give ourselves in addition to k, A and λ as above, an additive subgroup Λ of A such that $S_{-\lambda}(A) \subset \Lambda \subset S^{-\lambda}(A)$ and such that $\bar{a}\, x\, a \in \Lambda$ whenever $x \in \Lambda$ and $a \in A$. This done, we can associate to any quadratic module (M, [B]) the function $q_B : M \to A/\Lambda$, $q_B(m) \equiv B(m, m) \bmod \Lambda$, which depends only on [B] because $S_{-\lambda}(A) \subset \Lambda$. A <u>morphism</u> $\sigma : (M, [B]) \to (N, [C])$ is an A-linear map $M \to N$ such that for all $m, n \in M$, $\langle m, n\rangle_B = \langle \sigma m, \sigma n\rangle_C$ and $q_B(m) = q_C(\sigma m)$. If $\Lambda = S^{-\lambda}(A)$

(The maximum possible) the last restriction on σ is already implied by the other restrictions. The orthogonal sum $\sigma \perp \rho$ of two morphisms is defined in the obvious way. The automorphism group of $(M, [B])$ is denoted by $U(M, [B])$ and called the <u>unitary group</u> of $(M, [B])$ (or of M or of $[B]$). The full subcategory of non-singular quadratic modules is denoted by $\underline{\underline{Q}}^{\lambda,\Lambda}(A)$.

2. Hyperbolic Forms.

The <u>hyperbolic functor</u> $H : \underline{\underline{P}}(A) \to \underline{\underline{Q}}^{\lambda,\Lambda}(A)$, where $\underline{\underline{P}}(A)$ is the category of finitely generated projective right A-modules, is defined by $H(P) = (P \oplus P^*, [B_P])$ where $P^* = \mathrm{Hom}_A(P, A)$ is viewed, of course, as a right A-module and where $B_P((x, f), (y, g)) = f(y)$ for $x, y \in P$, $f, g \in P^*$. We shall write q_P and $\langle , \rangle_P$ in place of q_{B_P} and $\langle , \rangle_{B_P}$, respectively. Thus $q_P(x, f) \equiv f(x) \bmod \Lambda$ and $\langle (x, f), (y, g) \rangle_P = f(y) + \lambda \overline{g(x)}$. For a morphism $\alpha : P \to Q$ in $\underline{\underline{P}}(A)$ we put $H(\alpha) = \alpha \oplus \alpha^{*-1}$ which is easily seen to be a morphism in $\underline{\underline{Q}}^{\lambda,\Lambda}(A)$. The quadratic module $H(P)$ is called a <u>hyperbolic module</u>. There is a canonical isomorphism $H(P \oplus Q) \cong H(P) \perp H(Q)$. Moreover for any non-singular quadratic module $(M, [B])$ we have an isomorphism $(M, [B]) \perp (M, [-B]) \cong H(M)$.

We identify P^{**} with P, as usual, so that the dual $\alpha \mapsto \alpha^*$ defines a conjugate k-linear automorphism of order two of $\mathrm{Hom}_A(P^*, P)$ and similarly for $\mathrm{Hom}_A(P, P^*)$. Writing

$$\mathrm{End}_A(P \oplus P^*) = \begin{pmatrix} \mathrm{Hom}_A(P, P) & \mathrm{Hom}_A(P^*, P) \\ & \\ \mathrm{Hom}_A(P, P^*) & \mathrm{Hom}_A(P^*, P^*) \end{pmatrix}$$

an endomorphism $\sigma = \begin{pmatrix} \alpha & \beta \\ \gamma & \delta \end{pmatrix}$ belongs to U(H(P)) iff the following conditions hold:

1. $\begin{pmatrix} \alpha & \beta \\ \gamma & \delta \end{pmatrix}\begin{pmatrix} \delta^* & \lambda\beta^* \\ \lambda\gamma^* & \alpha^* \end{pmatrix} = I$ (σ preserves $\langle\,,\,\rangle_P$)

and

2. $((\alpha^*\gamma)(x))(x) \in \Lambda$ for all x in P, and

   $f((\beta^*\delta)(f)) \in \Lambda$ for all f in P* (σ preserves q_P)

If $\Lambda = S^{-\lambda}(A)$ (the maximum possible) then 2 is a consequence of 1.

If $\alpha \in \mathrm{Aut}_A(P)$ then $H(\alpha) = \begin{pmatrix} \alpha & 0 \\ 0 & \alpha^{*-1} \end{pmatrix}$.

A matrix $\begin{pmatrix} I & \beta \\ 0 & I \end{pmatrix}$ belongs to U(H(P)) iff

1. $\beta \in S^{-\lambda}(\mathrm{Hom}_A(P^*, P))$

and

2. $(\beta^* f)(f) \in \Lambda$ for all $f \in P^*$

Such β's form a subgroup $S^{-\lambda}(\mathrm{Hom}_A(P^*, P), \Lambda)$ of $S^{-\lambda}(\mathrm{Hom}_A(P^*, P))$ and we have a group monomorphism $E : S^{-\lambda}(\mathrm{Hom}_A(P^*, P), \Lambda) \to U(H(P))$, $E(\beta) = \begin{pmatrix} I & \beta \\ 0 & I \end{pmatrix}$, whose image is normalized H(Aut P). The action of the latter on the former corresponds to $\beta \mapsto \alpha^{-1}\beta(\alpha^{-1})^*$.

Let (M, [B]) be a quadratic module. Two elements x, y in M are said to form a <u>hyperbolic pair</u> if, for all a, b $\in$ A, $q_B(xa + yb) \equiv \bar{b}a \mod \Lambda$ and $\langle x, y\rangle_B = 1$. In this case x and y generate a subspace of (M, [B]) isomorphic to

H(A); such a subspace is called a <u>hyperbolic plane</u>. If P is a free A-module with basis $e_1, \ldots, e_n$ and dual basis $f_1, \ldots, f_n$ for P* then H(P) is the orthogonal sum of the hyperbolic planes spanned by the hyperbolic pairs $(e_1, f_1), \ldots, (e_n, f_n)$. Using the basis $e_1, \ldots, e_n, f_1, \ldots, f_n$ for $P \oplus P^*$ the unitary group U(H(P)) corresponds to the unitary group $\underline{U_{2n}(A)}$ of matrices $\sigma = \begin{pmatrix} \alpha & \beta \\ \gamma & \delta \end{pmatrix}$ such that

1. $\begin{pmatrix} \alpha & \beta \\ \gamma & \delta \end{pmatrix}\begin{pmatrix} \overline{\delta} & \lambda\overline{\beta} \\ \lambda\overline{\gamma} & \overline{\alpha} \end{pmatrix} = I$

and

2. $\overline{\gamma}\alpha$ and $\overline{\delta}\beta$ have diagonal entries in Λ.

Here the bar denotes the conjugate transpose of an element of $M_n(A)$. The maps $E : \beta \mapsto \begin{pmatrix} I & \beta \\ 0 & I \end{pmatrix}$ and $E' : \beta \mapsto \begin{pmatrix} I & 0 \\ \overline{\beta} & I \end{pmatrix}$ define homomorphisms into $U_{2n}(A)$ from the additive group $S_n^{-\lambda}(A, \Lambda)$ of those β in $S^{-\lambda}(M_n(A))$ whose diagonal entries lie in Λ.

3. Finiteness Conditions and Hyperbolic Rank.

The main theorems are proved under the following finiteness assumption:

(f) A is a finite k-algebra and the maximal ideal space max(k) of k is noetherian of finite dimension d.

We say an A-module P has <u>f-rank</u> ≥ r iff, for all $\underline{m}$ in max(k), the $A_{\underline{m}}$-module $P_{\underline{m}}$ has a free direct summand of rank r. A quadratic module (M, [B]) is said to have <u>hyperbolic rank</u> (or <u>h-rank</u>) ≥ r iff it has a subspace isomorphic to H(P) where f-rank P ≥ r.

4. Cancellation Theorem.

Throughout this section we assume the hypothesis (f) of section 3.

CANCELLATION THEOREM. Let (M, [B]), (M', [B']) and (N, [C]) be quadratic modules such that
$(M, [B]) \perp (N, [C]) \cong (M', [B']) \perp (N, [C])$. Then
$(M, [B]) \cong (M', [B'])$ provided that
(a) h-rank (M, [B]) > d
and
(b) (N, [C]) is non-singular (i.e. $(N, [C]) \in \underline{Q}^{\lambda,\Lambda}(A)$).

If (M, [B]) is a quadratic module we say an element x in M is unimodular in (M, [B]) iff there exists an element y in M such that $\langle y, x\rangle_B = 1$. If [B] is non-singular this definition coincides with the usual one, namely there exists an f in M* such that f(x) = 1.

A corollary and in fact a key step in the proof of the cancellation theorem is the following proposition.

PROPOSITION. If h-rank (M, [B]) > d + 1 then U(M, [B]) acts transitively on those unimodular elements in (M, [B]) of the same "length" (i.e. value under q_B).

We deduce the cancellation theorem from the proposition. It suffices to prove the theorem when (N, [C]) = H(A). For one can clearly replace (N, [C]) by $(N, [C]) \perp (N', [C'])$ for any non-singular (N', [C']); for a suitable choice of (N', [C']) one has $(N, [C]) \perp (N', [C']) \cong H(A^n)$. An induction on n then reduces the problem to the case $n = 1$.

Identify $(M', [B']) \perp H(A)$ with $(M, [B]) \perp H(A)$ via the given isomorphism and denote the image of $(M', [B'])$ by M' and the image of $H(A)$ by $H'(A)$. Note that

$$\text{h-rank}\big((M, [B]) \perp H(A)\big) = \text{h-rank}(M, [B]) + \text{h-rank}\, H(A) > d + 1.$$

If (x, y) and (x', y') are hyperbolic pairs which generate $H(A)$ and $H'(A)$ respectively the proposition therefore gives us a σ in $U((M, [B]) \perp H(A))$ such that $\sigma(x) = x'$. It is then not difficult to adjust σ so that in addition $\sigma(y) = y'$. Thus $\sigma(H(A)) = H'(A)$ and so

$$\sigma(M) = \sigma(\text{orthogonal complement } H(A)) = \text{orthogonal complement } H'(A) = M'.$$

This completes the proof.

A quadratic module $(N, [C])$ is said to <u>split</u> if $(N, [C]) = (U, [D]) \perp (V, [o])$ and $(U, [D])$ is non-singular, $(U, [D]) \perp (V, [o])$ is called a <u>splitting</u> of $(N, [C])$. If $(M, [B])$ is a quadratic module, a subspace $(N, [C])$ of $(M, [B])$ is called a <u>split subspace</u> if $(N, [C])$ is a split quadratic module and N is a direct summand of M.

As a corollary of the cancellation theorem we obtain Witt's theorem.

COROLLARY. <u>Let</u> $(M, [B])$ <u>and</u> $(M', [B'])$ <u>be non-singular quadratic modules such that there exists a morphism</u> $(M, [B]) \to (M', [B'])$. <u>Suppose</u> $(N, [C])$ <u>a split subspace of</u> $(M, [B])$, $(U, [D]) \perp (V, [o])$ <u>a splitting of</u> $(N, [C])$

and τ *any monomorphism*: $(N, [C]) \to (M', [B'])$ *such that* $\tau(N, [C])$ *is a split subspace of* $(M', [B'])$. *If (h-rank* $(M', [B'])$-(*f-rank* (V))-(*h-rank*$(U, [D])$))) $> d$ *then there exists a morphism* $\sigma : (M, [B]) \to (M', [B'])$ *extending* τ.

5. Normal subgroups of $U_{2n}(A)$.

For the moment we drop the hypothesis (f) on A. The letter $\underline{q}$ always denotes a two-sided involution invariant ideal of A. The letter R when it appears with $\underline{q}$ in the manner $(\underline{q}, R)$ always denotes a subgroup of $\Lambda \cap \underline{q}$ with the following properties:

(1) If $q \in \underline{q}$ and $x \in \Lambda$ then $q - \lambda\bar{q}$ and $\bar{q}xq$ lie in R

(2) If $a \in A$ and $r \in R$ then $\bar{a}ra \in R$.

Recall that the involution on A extends to $M_n(A)$ by setting $\bar{\alpha}$=conjugate transpose of α. Define $\underline{S_n^{-\lambda}(\underline{q}, R)}$ to be the subgroup of all elements of $S^{-\lambda}(M_n(\underline{q}))$ whose diagonal entries lie in R. The $S_n^{-\lambda}(\underline{q}, R)$ are invariant under the action of $GL_n(A)$ on $S_n^{-\lambda}(A, \Lambda)$ defined by $\alpha \mapsto \bar{\rho}\alpha\rho$. In fact if on the couple $(\underline{q}, R)$ we drop the restriction that $R \supset \{\bar{q}xq : x \text{ in } \Lambda \text{ and } q \text{ in } \underline{q}\}$ then the $S_n^{-\lambda}(\underline{q}, R)$ are precisely all the $GL_n(A)$-invariant subgroups of $S_n^{-\lambda}(A, \Lambda)$.

Recall the maps E and E' at the end of section two. We write $\underline{EU_{2n}(A)}$ for the subgroup of $U_{2n}(A)$ generated by the images of E and E' and $\underline{EU_{2n}(\underline{q}, R)}$ for the normal subgroup of $EU_{2n}(A)$ generated by $E(S_n^{-\lambda}(\underline{q}, R))$ and $E'(S_n^{-\lambda}(\underline{q}, R)$. It may be possible for a fixed $\underline{q}$ that several R's define the same "EU-subgroup" of $U_{2n}(A)$ and so in writing $EU_{2n}(\underline{q}, R)$

we insist R stand for the maximum such R, which exists.

For the remainder of this section we invoke the hypothesis (f) of section three.

LEMMA. *If* $n > d + 2$ *then* $EU_{2n}(\underline{q}, R)$ *is normal in* $U_{2n}(A)$.

Assume $n > d + 2$ and define *$FU'_{2n}(\underline{q}, R)$* to be that subgroup of $U_{2n}(A)$ which centralizes $EU_{2n}(A)$ mod $EU_{2n}(\underline{q}, R)$. The canonical homomorphism $A \to A/\underline{q}$ induces a homomorphism $U_{2n}(A) \to U_{2n}(A/\underline{q})$ the kernel of which we denote by *$U_{2n}(\underline{q})$*. If *$U'_{2n}(\underline{q})$* denotes the inverse image of the centre of $U_{2n}(A/\underline{q})$ it is not difficult to check that $FU'_{2n}(\underline{q}, \Lambda \cap \underline{q}) = U'_{2n}(\underline{q})$. We define *$FU_{2n}(\underline{q}, R)$* $= FU'_{2n}(\underline{q}, R) \cap U_{2n}(\underline{q})$. It follows that $FU_{2n}(\underline{q}, \Lambda \cap \underline{q}) = U_{2n}(\underline{q})$.

THEOREM. *If* $n > d + 2$ *a subgroup H of* $U_{2n}(A)$ *is normalized by* $EU_{2n}(A)$ *iff for some* $(\underline{q}, R)$ $EU_{2n}(\underline{q}, R) \subset H \subset FU'_{2n}(\underline{q}, R)$. *Moreover* $\underline{q}$ *and* R *are unique*.

If H lies between $EU_{2n}(\underline{q}, R)$ and $FU'_{2n}(\underline{q}, R)$ we say H is of *level* $(\underline{q}, R)$.

We shall sketch a proof based on the following two lemmas.

LEMMA 1. *Suppose H is normalized by* $EU_{2n}(A)$ *and* $EU_{2n}(\underline{q}, R) \subset H \subset U_{2n}(\underline{q})$. *If* $n > d + 2$ *the following are* equivalent:

(a) *R is maximum with respect to the inclusion*

$EU_{2n}(\underline{q}, R) \subset H$

(b) $H \subset FU_{2n}(\underline{q}, R)$

LEMMA 2. <u>Suppose H is normalized by</u> $EU_{2n}(A)$ and $\underline{q}$ <u>and</u> H <u>are maximum such that</u> $EU_{2n}(\underline{q}, R) \subset H$. <u>If</u> $n > d + 2$ <u>then</u> $H \subset U'_{2n}(\underline{q})$.

We sketch a proof of the theorem.

Choose $\underline{q}$ and R as in lemma 2. Since $n > 2$,

$$\begin{aligned} EU_{2n}(\underline{q}, R) &= [EU_{2n}(\underline{q}, R), EU_{2n}(A)] \text{ (mixed commutator)} \\ &\subset [H, EU_{2n}(A)] \\ &= H_o \subset [U'_{2n}(\underline{q}), EU_{2n}(A)] \\ &\subset U_{2n}(\underline{q}). \end{aligned}$$

R is again maximum with respect to the inclusion $EU_{2n}(\underline{q}, R) \subset H_o$ and so lemma 1 implies $H_o \subset FU_{2n}(\underline{q}, R)$. Therefore $[H_o, EU_{2n}(A)] \subset EU_{2n}(\underline{q}, R)$ and $[[H , EU_{2n}(A)], EU_{2n}(A)] \subset EU_{2n}(\underline{q}, R)$. (2. Chapter V, lemma (5.1)) implies $[H, EU_{2n}(A)] \subset EU_{2n}(\underline{q}, R)$ and so by definition $H \subset FU'_{2n}(\underline{q}, R)$.

The uniqueness part of the theorem follows from the observation that

$EU_{2n}(\underline{q}, R) = [EU_{2n}(\underline{q}, R), EU_{2n}(A)] = [FU'_{2n}(\underline{q}, R), EU_{2n}(A)]$ implies $EU_{2n}(\underline{q}, R) = [H, EU_{2n}(A)]$. If H is both of level $(\underline{q}, R)$ and level $(\underline{q}', R')$ then $EU_{2n}(\underline{q}, R) = EU_{2n}(\underline{q}', R')$ and this implies $\underline{q} = \underline{q}'$ and $R = R'$.

6. Symplectic Case.

Assume A has trivial involution (thus A is commutative), $\lambda = -1$ and $\Lambda = S^{-\lambda}(A) = A$. Write Sp in place of U and XSp in place of XU where X is any letter. We call a quadratic module a symplectic module and $Sp_{2n}(A)$ a symplectic group of A. We obtain stronger versions of our theorems.

CANCELLATION THEOREM. *Let* (M, [B]), (M', [B']) *and* (N, [C]) *be symplectic modules such that*

$(M, [B]) \perp (N, [C]) \cong (M', [B']) \perp (N, [C])$. Then

$$(M, [B]) \cong (M', [B']) \text{ provided that}$$

(a) (M, [B]) *contains a non-singular subspace* (P, [D]) *and f-rank* P $\geq$ d

(b) (N, [C]) is *non-singular*.

The theorem was originally demonstrated by Bass in the case (M, [B]) is non singular.

THEOREM. *If* $2n > \max(d + 1, 4)$ *a subgroup* H *of* $Sp_{2n}(A)$ *is normalized by* $ESp_{2n}(A)$ *iff, for some* $(\underline{q}, R)$, $ESp_{2n}(\underline{q}, R) \subset H \subset FSp'_{2n}(\underline{q}, R)$. *Moreover* $\underline{q}$ *and* R *are unique and squares of elements in* $\underline{q}$ *lie in* R.

7. Some Corollaries.

We invoke the hypothesis (f) of section three. For a fixed λ and Λ define $KU_1^\Lambda(\underline{q}, R) = \varinjlim_n FU_{2n}(\underline{q}, R)/EU_{2n}(\underline{q}, R)$ and write $KU_1^\Lambda(A)$ in place of $KU_1^\Lambda(A, \Lambda)$.

COROLLARY 1. $KU_1^\Lambda(q, R)$ *is abelian*.

COROLLARY 2. *If* $n > d + 2$ *the natural homomorphism*

$FU_{2(d+1)}(\underline{q}, R) \to FU_{2n}(\underline{q}, R)/EU_{2n}(\underline{q}, R)$ is surjective.

COROLLARY 3. If A is an order in a finite semi-simple $\mathbb{Q}$-algebra then $KU_1^{\Lambda}(A)$ is finitely generated.

COROLLARY 4. If A is semi-local then the natural homomorphism $KU_1^{\Lambda}(A) \to KU_1^{S^{-\lambda}(A)}(A)$ is surjective.

REFERENCES

1. E. Artin Geometric Algebra, Interscience, No.3 (1957)
2. H. Bass Algebraic K-Theory, Benjamin, (1968).

Institut des Hautes Études Scientifiques.

NORMAL SUBGROUPS OF INTEGRAL ORTHOGONAL GROUPS

by Martin Kneser

Let K be an algebraic number field, S a finite set of places of K containing all archimedean places, and I the ring of elements in K which are in the valuation ring of v for all $v \notin S$. In particular, if S contains only the archimedean places, I is the ordinary ring of integers in K. For a linear K - algebraic group G, considered in a fixed embedding $G \subseteq GL_n$, an S - congruence subgroup of G(I) is a subgroup containing all matrices $x \equiv 1 \bmod \mathfrak{m}$ in G(I) for some non - zero ideal $\mathfrak{m}$ of I. These S-congruence subgroups have finite index in G(I). The congruence subgroup problem is the question whether conversely every subgroup of finite index is an S - congruence subgroup. If G is semi-simple and connected, and G(I) is not finite, it is easily seen that this can be true only if G is simply connected, cf [6]. We therefore restrict our attention to simply connected semi-simple groups. The congruence subgroup problem has been solved for Chevalley groups (cf [1, 3,4,6,7]), and in [8], some results for Spin groups of quadratic forms of Witt index ≥ 2 are given.

Since every subgroup of finite index contains a normal subgroup of finite index it suffices to consider these. In fact, one can get information about all normal subgroups of G(I). In this note we announce some partial results on normal subgroups of G(I) for Spin groups of quadratic forms of arbitrary Witt index (including index 0). There seems to be no essential difficulty to prove theorems similar to the one below for other classical groups.

Let X be a vector space of finite dimension n over K with a non - degenerate quadratic form q, i(X) its Witt index, i.e. the dimension of a maximal isotropic subspace of X, and $i_S(X) = \sum_{v \in S} i\,(X \otimes K_v)$. For any subset Y or X, denote by $Y^{\perp}$ the subspace of all vectors in X, orthogonal to Y. Let G be the Spin group of (X,q), operating on X in the natural way, and H_n an S - congruence subgroup of G(I). On K we introduce the S - congruence topology with the non-zero ideals of I as a basis of neighbourhoods of 0. This induces a topology on the set of rational points of any affine K - algebraic variety. All our results depend on the following crucial

LEMMA. Let $x \in X$, $q(x) \neq 0$, $Kx^{\perp} = Y$, and N a non-central normal subgroup of H_n. Assume $n \geq 5$, and

(1) $i_S(X) \geq 2$, $i_S(Y) \geq 1$.

Then the map $g \mapsto gx$ of N into G(K)x is open in the S - congruence topology. If H_{n-1} is an S - congruence subgroup of the Spin group of Y, considered as a subgroup of G(K), then $N \cap H_{n-1}$ is a non - central subgroup of H_{n-1}.

The proof is technical and uses ideas of [2] together with strong approximation in the Spin group of Y.

CONSEQUENCES. i) NH_{n-1} is an S - congruence subgroup. By induction on n - m we get

ii) Let X be the orthogonal direct sum of a subspace Y of dimension $m \geq 4$ and another subspace Z, assume (1), and let H_m be an S - congruence subgroup of the Spin group of Y. Then $H_m N$ is an S - congruence subgroup.

iii) Under the assumptions of ii), the smallest normal subgroup L of H_n containing H_m is an S - congruence subgroup.

THEOREM. Assume $\dim X = n \geq 8$, $i_S(X) \geq 2$, and N a non - central normal subgroup of H_n. Then there exists an S - congruence subgroup L, depending only on H_n (not on N), and another one M, depending on N, such that N contains the mixed commutator subgroup [L,M].

PROOF. Write X as an orthogonal direct sum as in ii) with $i_S(Y) \geq 1$, $i_S(Z) \geq 1$, and define L as in iii). Let H_{n-m} be an S - congruence subgroup of the Spin group of Z and M a normal S - congruence subgroup contained in $H_{n-m}N$. Since H_m and H_{n-m} commute, H_m and M commute modulo N. Since M and N are normal, all the conjugates of H_m commute with M modulo N, and so $[L,M] \subseteq N$.

Let us finally translate some of these results into the language of group extensions (cf [1,3,5,6]).
On G(K) we consider two topologies: The S - congruence topology with the S - congruence subgroups as basis of neighbourhoods of 1, and the S - arithmetic topology with the

subgroups of $G(I)$ of finite index as basis. Let $\overline{G}(K)$ resp. $\hat{G}(K)$ be the completion of $G(K)$ in these topologies. The identity mapping of $G(K)$ induces a homomorphism $\hat{G}(K) \to \overline{G}(K)$, which is surjective and has a certain kernel $C^S(G)$. This kernel is $\{1\}$ if and only if every subgroup of $G(I)$ of finite index is an S-congruence subgroup.

The theorem above, together with [2], implies that $C^S(G)$ is contained in the centre of $\hat{G}(K)$, if $\dim X \geq 8$, $i_S(X) \geq 2$.

Under the assumptions of ii), denote by G_n resp. G_m the Spin group of X resp. Y. The inclusion $G_m \to G_n$ induces a homomorphism $C^S(G_m) \to C^S(G_n)$, and ii) implies that this is surjective. Now $C^S(G)$ is known [3] whenever $n \geq 5$ and X has maximal Witt index, namely with $\mu(K)$ the group of roots of unity contained in K, we have

$$(2) \quad C^S(G) = \begin{cases} 1 & \text{if S contains a real or a non-archimedean place} \\ \mu(K) & \text{if S contains only complex places} \end{cases}$$

If $i(X) \geq 2$, X contains a five-dimensional space of maximal Witt index, and is contained in such a space of dimension 2n. So, by the argument above, (2) holds in this case too (cf [8]). If $i(X) = 1$, Serre's results [7] about SL_2 and the isomorphisms of four-dimensional Spin groups may be used to get the following:

$C^S(G) = \{1\}$ if $i(X) = 1$, and S contains a real place v with $i(X \otimes K_v) \geq 2$, or a non-archimedean place;

$c^S(G) = \mu(K)$ if $i(X) = 1$ and S contains only complex places,

$c^S(G)$ is finite if $i(X) = 1$, $i_S(X) \geq 2$.

References

1. H. Bass, J. Milnor and J-P. Serre, Solution of the Congruence Subgroup Problem for SL_n $(n \geq 3)$ and Sp_{2n} $(n \geq 2)$, Publ. Math. I.H.E.S. no 33 (1968) 59-137

2. M. Kneser, Orthogonale Gruppen über algebraischen Zahlkörpern, J. reine u. angew. Math. 196 (1956) 213-220

3. H. Matsumoto, Sur les sous-groupes arithmétiques des groupes semi-simples déployés, Ann. Scient. Éc. Norm. Sup., 4ème serie, t.2, (1969) 1-62

4. J. Mennicke, Finite factor groups of the unimodular group, Annals of Math. 81(1965) 31-37

5. C. Moore, Group extensions of p-adic and adelic linear groups, Publ. Math. I.H.E.S. No.39, (1969) 5-74

6. J-P. Serre, Groupes de congruence, Sém. Bourbaki 1966/67, exp. 330

7. J-P. Serre, Le problème des groupes de congruence pour SL_2,

8. L. I. Vaserstein, Subgroups of finite index in Spin groups of rank ≥ 2 (Russian), Mat. Sbornik 75(1968) 178-184

University of Göttingen.

A SPLITTING THEOREM AND THE KÜNNETH FORMULA IN ALGEBRAIC K-THEORY

W.C. HSIANG

(Notes by C.B. Thomas)

I shall summarize some joint work with F.T. Farrell [1], [2], [3] in this lecture.

Let $\pi = G \times_\alpha Z$ be a split extension of G by Z, where α is the automorphism of G induced by the action of the generator t of Z. Write A for ZG, the integral group ring of G. It follows that

$$Z(G\times_\alpha Z) = A_\alpha[t, t^{-1}],$$

the α-twisted finite Laurent series ring over A, that is, if at^m, bt^n are two monomials over A, $(at^m)(bt^n) = a\alpha^{-m}(b)t^{m+n}$.

Let $C(A, \alpha)$ be the Grothendieck group defined by pairs (P, ν), P a finitely generated projective A-module and ν an α-linear nilpotent endomorphism of P. The map $P \to (P, 0)$ embeds K_0A as a direct summand of $C(A, \alpha)$; write $\mathrm{Nil}(A, \alpha)$ for the complement.

<u>THEOREM 1</u>, see [2], [3]. $\mathrm{Wh}(G\times_\alpha Z) = \tilde{X} \oplus \mathrm{Nil}(A,\alpha) \oplus \mathrm{Nil}(A,\alpha^{-1})$, <u>where $\tilde{X}$ is defined by the short exact sequence</u>

$$0 \to \mathrm{Wh}(G)/_{I(\alpha)} \to \tilde{X} \to (\tilde{K}_0 G)^{\alpha_*} \to 0.$$

$I(\alpha) = \{x : x = y - \alpha_* y,\ y \in \mathrm{Wh}(G)\}$, and the superscript α_* refers to α invariant elements.

REMARKS: 1. There is a general formula for $K_1A_\alpha[t, t^{-1}]$ like Theorem 1 for any A, and $Nil(A, \alpha^{-1}) \cong Nil(A, \alpha)$ for a ring A with involution.

2. The corresponding short exact sequence in the formula for $K_1A_\alpha[t, t^{-1}]$ does not in general split.

3. Inductively one can use this result to prove that $Wh(\pi_1$ surface$) = 0$, and that $Wh(\pi_1$ solvmanifold$) = 0$. (One knows also that $\tilde{K}_0(\pi_1$ solvmanifold$) = 0$.)

Let p be the composition

$$Wh(G\times_\alpha Z) \underset{proj}{\longrightarrow} Nil(A,\alpha) \oplus \tilde{K}_0(G)^{\alpha_*} \underset{incl}{\hookrightarrow} Nil(A,\alpha) \oplus \tilde{K}_0G.$$

In the geometric applications let M^n, M'^n $(n \geq 6)$ be compact (closed) smooth or piecewise linear manifolds. $f : M^n \to M'^n$ is to be a fixed homotopy equivalence, and both fundamental groups are identified with $G\times_\alpha Z$, via f_*. Assume also that G is finitely presented. Suppose that M'^n contains a submanifold N'^{n-1} with fundamental group G. If there exists a submanifold N^{n-1} of M^n and a homotopy equivalence of pairs

$$g : M^n, N^{n-1} \to M'^n, N'^{n-1}$$

such that $g \simeq f$ on M^n, we shall say that f <u>splits</u> along N^{n-1}.

THEOREM 2, (announced in [2]). *The obstruction O(f) to splitting f equals* $p\tau(f)$, *where* $\tau(f)$ *is the torsion of the homotopy equivalence f.*

Idea of proof. Without loss of generality suppose f transverse regular, and let N be the inverse image of N'.

In the regular covering space of M with fundamental group G, N(N') and its translates define left and right components A, B (A', B'), as illustrated.

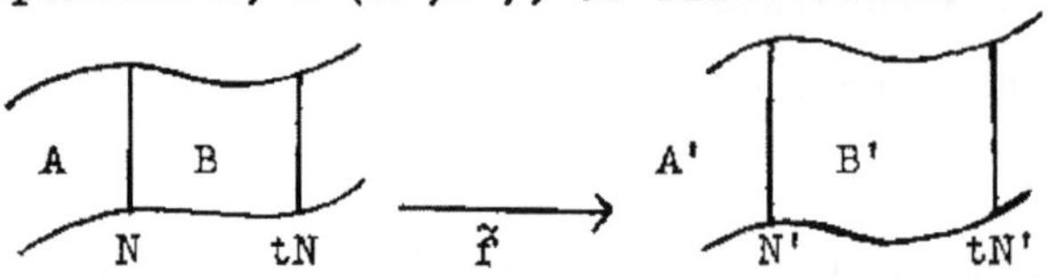

We now attempt to replace N by a submanifold homotopy equivalent to N' by handle exchanges between A and B. These must be done for each translate of N inside its adjacent "boxes", that is, Z-equivariantly, so that at each stage we may drop down to M.

In order to illustrate the obstruction let M have dimension 2r+1. The construction proceeds until we reach the last homology kernel $K_r(N;\ ZG)$, which by the Mayer-Vietoris theorem splits as a direct sum
$K_r(N;\ ZG) = K_{r+1}(A,\ N;\ ZG) \oplus K_{r+1}(B,\ N;\ ZG)$.
The first summand is projective (so is the second), and via the covering transformation t, defines the obstruction (P, t) in C(A, α). If this is zero, we may pick a basis for the (stably) free module $K_{r+1}(A,\ N;\ ZG)$, which behaves nicely under translation, and complete the handle exchange.

A first application of this theorem is to the existence of topologically non-trivial h-cobordisms, see [1]. If $L^3_{p^2}$ is a lens space with fundamental group Z_{p^2}, there exist h-cobordisms W^{n+1} between $M'^n = S^1 \times T^{n-4} \times L^3_{p^2}$ and M^n, such that the torsion of the homotopy equivalence $f : M^n \to M'^n$ projects to a non-zero element in $\mathrm{Nil}(Z(Z^{n-4} \times Z_{p^2}), 1)$ but to zero in $\tilde{K}_o(Z^{n-4} \times Z_{p^2})$.

If there were a homeomorphism $h : M^n \to M'^n$, the pull-back of the open submanifold $N'^{n-1} \times (-1, 1)$ would split in the G-covering space $\tilde{M}^n$, contradicting the choice of $O(f)$.

REMARKS: 1. It is enough to assume that M'^n is a Poincaré complex.

2. For the relative version of the theorem one supposes that the boundary is already split.

HIGHER INDEX OF HIRZEBRUCH-NOVIKOV

Now, let me describe an unpublished joint result with F.T. Farrell.

Let M^n be such that $H_1(M^n, Z)$ is of rank l. We define the higher index as a homomorphism

$$I : \Lambda^t \operatorname{Hom}(H_1, Z) \to Z, \quad n - t = 4k,$$

as follows. Suppose first that $l < n$. Choose a basis $\underset{\sim}{x}_1, \ldots, \underset{\sim}{x}_l$ of $\operatorname{Hom}(H_1, Z)$ in such a way that the dual classes are represented by submanifolds $N_i^{n-1} \subset M^n$ in general position, $i = 1, 2, \ldots, l$. If

$$N^{4k} = \bigcap_{i=1}^{t} N_i^{n-1}, \quad I(\underset{\sim}{x}_{i_1} \wedge \ldots \wedge \underset{\sim}{x}_{i_t}) = \operatorname{Index}(N^{4k}).$$

The index of the manifold $I(M)$ can then be written as the sum

$$\sum_{i,t} I(\underset{\sim}{x}_{i_1} \wedge \ldots \wedge \underset{\sim}{x}_{i_t}) \cdot \underset{\sim}{x}_{i_1} \wedge \ldots \wedge \underset{\sim}{x}_{i_t}.$$

In general, we define $I(M) = I(M \times \mathbf{CP}^2)$.

THEOREM 3. In the piecewise linear category $I(M)$ is an invariant of the homotopy type of M.

The proof is by splitting, and uses the relative version of Theorem 2 above. The cases $l = 1$, $l = 2$ were proved earlier by Novikov and Rohlin respectively.

THEOREM 4. (Shaneson-Wall [4]) *Let* N^n, $n \geq 6$, *have fundamental group* G *and orientation map* $\omega : G \to Z_2$. *Let* ω_1 *be the composite of* ω *and the projection of* $G \times Z$ *onto* G. *Then there is a split short exact sequence of surgery obstruction groups*

$$0 \to L_{n+1}(G,\omega) \xrightarrow{\beta(N)} L_{n+1}(G \times Z, \omega_1) \xrightarrow{\alpha(N)} L_n^h(G,\omega) \to 0.$$

Where $L_i^h(\ ,\ )$ is Wall's obstruction group for homotopy equivalence and $L_i(\ ,\ )$ is that for simple homotopy equivalence.

Idea of proof. Consider the same situation as before, only with the addition of stable normal bundles, that is, we are given $f : M^{n+1} \to N^n \times S^1$ representing a class in the **bordism** group $\Omega_{n+1}(N^n \times S^1, \nu)$. By definition of the absolute groups $L_{n+1}(G \times Z, \omega_1)$ for manifolds with boundary, $f|\partial M$ is a simple homotopy equivalence, which splits by Theorem 2. Hence by restriction to the inverse image of $N \times *$, f defines an element of $\Omega_n(N^n, \nu)$. Choose f to represent an arbitrary class θ of $L_{n+1}(G \times Z, \omega_1)$, and define $\alpha(N)\theta$ to be the obstruction to replacing the restriction of f by a homotopy equivalence. It follows from Theorem 1 that $\alpha(N)$ is well-defined; in particular, if $\theta = 0$, f can be replaced by a split homotopy equivalence and $\alpha(N)\theta = 0$. The left inverse for $\alpha(N)$ is defined by taking products with S^1.

In order to see how $\beta(N)$ arises, suppose that $\alpha(N)\theta = 0$. This implies that f is bordant to $g : P^{n+1} \to N^n \times S^1$ such that g restricted to $Q^n = g^{-1}(N^n \times *)$ is a homotopy equivalence. Cutting P along Q now yields an element in the relative group $\Omega_{n+1}(N^n \times I, \nu)$, which in turn defines an element of $L_{n+1}(G, \omega)$.

Theorem 4 applies in particular to free abelian groups. The split short exact sequence

$$0 \to L_{n+1}(Z^{l-1}) \to L_{n+1}(Z^l) \to L_n(Z^{l-1}) \to 0$$

enables us to compute the obstruction groups for Z^l by induction from the known result for $l = 0$.

THEOREM 5. *If* $f : M \to M'$ *induces a map of stable normal bundles, then* $I(M) - I(M')$ *coincides with the index part of the Wall surgery obstruction* $L(f) \in L_n(Z^l)$.

Loosely speaking, the difference between the indices measures the infinite part of the difficulty in replacing f by a homotopy equivalence.

REFERENCES

1. F.T. Farrell and W.C. Hsiang, *H-cobordant manifolds are not necessarily homeomorphic*, Bull.Amer.Math.Soc. 73 (1967), 741-744.
2. F.T. Farrell and W.C. Hsiang, *A geometric interpretation of the Kunneth formula in algebraic K-theory*, Bull. Amer.Math.Soc. 74 (1968), 548-553.
3. F.T. Farrell and W.C. Hsiang, *A formula for* $K_1R_\alpha[T]$, Categorical Algebra, Symp. of Pure Math. A.M.S. (to appear).
4. J. Shaneson, *Wall's surgery obstruction groups for* $Z \times G$, Bull.Amer.Math.Soc. 74 (1968), 467-471.

OBSTRUCTIONS FOR GROUP ACTIONS ON S^{2n-1}

C.B. THOMAS

Suppose that the finite group π operates freely on S^{2n-1}, then a simple argument with the spectral sequence of a covering shows that π has periodic cohomology with period dividing 2n. However this necessary condition for a free action is known not to be sufficient. The simplest counter example is $\mathfrak{S}_3$, the symmetric group on three letters, which has elements outside its centre of order 2. It follows by a theorem of Milnor [2], that $\mathfrak{S}_3$ cannot operate freely on S^{2n-1}, for any value of n. The purpose of this lecture is to sketch an obstruction theory for producing a π-action, starting with no restriction other than periodic cohomology.

If one relaxes the condition that the total space is a manifold, that is, one asks only for a free π-action on a finite CW-complex homotopy equivalent to the sphere, the periodic cohomology condition becomes sufficient.

THEOREM 1. (Swan [4]) Let π have order r and cohomological period 2k, and write $d = (r, \varphi(r))$, where φ is the Euler function. There exists a finite simplicial complex Σ^{2n-1}, $n = kd$, on which π acts freely and simplicially.

Swan's method of proof is to construct a $Z\pi$-resolution for Z of the form

$$0 \to Z \to F_{2k-1} \to P_{2k-2} \to \cdots \to F_o \to Z \to 0, \qquad (1)$$

where P_{2k-2} is the only non-free projective module.

P_{2k-2} measures the obstruction in $K_0\pi$ to realising (1) by a finite complex, and may be killed by splicing together d copies of (1). This argument has two important consequences:

(a) The quotient complex $Y = \Sigma^{2n-1}/_{\pi}$ is orientable and satisfies Poincaré duality, and

(b) By a single further splice, which kills an obstruction in $K_1\pi$, one can ensure that the chain homotopy equivalence $C_*Y \to C^*Y$ is simple.

The next part of the programme is to construct a vector bundle over Y with spherical Thom class; for reasons that will soon become obvious we work in the piecewise linear category.

Since Y satisfies Poincaré duality, there is an essentially unique S^{N-1} fibration ($N \gg n$), the Spivak fibration [3], over Y with spherical Thom class. The problem is to reduce the stable structure group of this fibration from F to PL.

THEOREM 2. (Wall-Thomas) <u>If</u> π <u>is soluble, the Spivak fibration</u> ν <u>over</u> Y <u>is fibre homotopically equivalent to a PL-bundle</u>.

Sketch proof. The obstructions to reducing the group are the obstructions to a cross-section of the fibration ξ associated to ν with fibre F/PL. Since Y is the (2n-1) skeleton of an Eilenberg-MacLane complex $K(\pi, 1)$, the obstructions lie in the groups

$H^p(\pi, \pi_{p-1}(F/PL))$, $p \leq 2n-2$, and $H^{2n-1}(Y, \pi_{2n-2}(F/PL))$. However, if all the lower dimensional obstructions vanish, so does the top one. To see this one only has to note that the group of the fibration has already been reduced outside some D^{2n-1}, and that the obstruction element $\varphi : S^{2n-2} \to F/PL$ bounds in $\Omega^{Poin}_{2n-2}(F/PL)$. One then appeals to

LEMMA A. <u>The Hurewicz map</u> $\pi_{2n-2}(F/PL) \to \Omega^{Poin}_{2n-2}(F/PL)$ <u>is injective</u>.

It follows that it is enough to consider the lower dimensional obstructions, all of which lie in the cohomology of π. Because of the periodic condition the only possible non-vanishing obstructions lie in

$$H^{4q-1}(\pi, \pi_{4q-2}(F/PL)) = H^{4q-1}(\pi, Z_2).$$

Note that this shows immediately that the group of ν reduces when π has odd order. When 2 divides $(\pi : 1)$ one has to appeal to a particular case of a lemma proved by E. Thomas:

LEMMA B. <u>Let</u> $\rho : Y_2 \to Y$ <u>be the projection of a covering complex for</u> Y <u>with fundamental group a 2-Sylow subgroup</u> π_2, <u>and suppose that the induced fibration</u> $\rho^! \xi$ <u>admits a cross-section</u> s_2. <u>Then</u> $s_2 = \rho^! s$ <u>for some section</u> s <u>of</u> ξ <u>provided that for all</u> p,

(i) $H^p(\pi;\pi_{p-1}(F/PL)) \xrightarrow{\rho^*} H^p(\pi_2;\pi_{p-1}(F/PL))$ <u>is injective</u>, <u>and</u>

(ii) $H^p(\pi;\pi_p(F/PL)) \xrightarrow{\rho^*} H^p(\pi_2;\pi_p(F/PL))$ <u>is surjective</u>.

$$\begin{array}{ccc} \rho^{!}\xi & \dashrightarrow & \xi \leftarrow F/PL \\ \downarrow\uparrow s_2 & & \downarrow\uparrow s \\ Y_2 & \xrightarrow{\rho} & Y \end{array}$$

The hypotheses of this lemma are satisfied, whenever π is such that $H_1(\pi;\ Z_2) \cong H_1(\pi_2;\ Z_2)$. Because π has periodic cohomology, π_2 is either cyclic or generalised quaternionic, and in either case Y_2 is homotopically equivalent to a manifold. (There are enough orthogonal π_2 actions on S^{2n-1} to cover all possible k-invariants.) Hence s_2 exists. Condition (i) above is always satisfied; condition (ii) depends on the H_1 assumption, and the fact that the two period of π divides 4.

The soluble groups with periodic cohomology, which do not satisfy the H_1 assumption, or for which the obstructions do not vanish trivially, all have a generalised polyhedral subgroup P_v^* containing π_2, see page 179 in [6]. Lemma B applies to these groups with P_v^* replacing π_2; there are a few more technicalities, notably the computation of the 3-period of π.

DEFINITION. A specific reduction of the group of ν to PL is called a <u>normal invariant</u> of the complex Y.

Theorem 2 shows that, at least when π is soluble, normal invariants for Y exist (they are classified by [Y : F/PL]), and hence that it is reasonable to try and replace Y with a manifold by surgery. The orientation homomorphism is trivial, and the middle dimensional obstruction θ lies in $L_{2n-1}(\pi)$, see [5], which is closely related to the special unitary analogue of K_1. Furthermore the existence of the lens spaces shows that, for a <u>suitably chosen</u> reduction of ν, surgery is effective when $\pi \cong Z_p$. This suggests that

(a) we compare the obstruction θ to the lifted obstructions $\rho^*\theta$ for the covering spaces of Y with cyclic fundamental groups,

(b) we choose a normal invariant for Y, which lifts to one equivalent to the normal invariant of a manifold for this class of coverings, and

(c) we investigate the effect on θ of varying the (simple) homotopy type of Y.

Algebraically step (a) corresponds to considering L_{2n-1} as a contravariant functor, satisfying Frobenius reciprocity, with domain the category $\mathcal{S}_\pi$ of subgroups of π and inclusion maps. $I^* : L_{2n-1}(\pi) \to L_{2n-1}(\pi')$ is defined by restricting the action of π on the standard kernel to π', and the transfer I_* by the tensor product $\otimes_{\pi'} Z_\pi$. As yet the general theory is in a fragmentary state, and in particular I have no estimate of the exponent of the subgroup

$$K^C(\pi) = \bigcap_{\pi' \text{cyclic}} \operatorname{Ker} I^*.$$

The following result is an example of what can be done towards step (b). However, even for the small class of groups considered, the argument is **incomplete**, since a normal invariant ν consists of a bundle η plus a homotopy class in $\pi_{N+2n-1}(Y^{\eta})$. In choosing η nicely, we may loose control over the class, which in turn may affect the lifted surgery obstruction. Our argument also shows that steps (b) and (c) are interdependent.

THEOREM 3. *Let* p, q *be distinct primes*, $2 \leqslant q < p$, *and let* π *have presentation*
$\pi = \{x, y : x^p = y^q = 1, y^{-1}xy = x^r, r^q \equiv 1 \bmod p\}$.
There exists a normal invariant ν *for a Swan complex* Y_{π}^{4n-1}, *the bundle of which lifts to a bundle equivalent to the stable normal bundle of a lens space for the covering spaces with fundamental groups generated by* x *and* y.

Sketch proof. Note that, provided step (b) can be completed, it is enough for (a) to consider the subgroups generated by x and y. Any other cyclic subgroup is conjugate to $\{y\}$, and surgery obstructions are invariant under inner automorphisms. Let Y_p and Y_q be the corresponding covering complexes; the Atiyah-Hirzebruch spectral sequence $\hat{H}^p(\ ; K^q(pt)) \Rightarrow \hat{K}^*$ collapses for Y, Y_p and Y_q. Furthermore π has q-period 2 and p-period 2q, so in terms of graded groups one has

$$\bigoplus_{2n-1} Z_q + \bigoplus_{\frac{2n}{q}-1} Z_p \quad \begin{matrix} \nearrow G\tilde{K}^\circ(Y_p) \cong \bigoplus_{2n-1} Z_p \\ \searrow G\tilde{K}^\circ(Y_q) \cong \bigoplus_{2n-1} Z_q \end{matrix}$$

The maps are defined by restriction; the upper is monic on p-torsion and the lower on q-torsion, since x and y generate Sylow subgroups. By varying Swan's original construction of the complex Y^{4n-1} we can realise all possible homotopy types, and in particular can suppose that

$$Y_p \simeq L^{4n-1}(p; \underbrace{q, \ldots, q}_{2n}).$$

If $1 + \sigma$ is the complex line bundle over the lens space $L^{4n-1}(p; 1, \ldots, 1)$ associated to the defining representation, and R denotes the filtration on K° defined by the spectral sequence, $R_{2q}(Y_p)/R_{2q+1}(Y_p)$ is a cyclic group of order p, generated by σ^q. (This computation can be extracted from [1].) It follows from the diagram of graded groups, that some p-torsion element in $\tilde{K}^\circ Y$ maps under restriction to $-(2n)\sigma^q$, which defines the stable normal bundle of $L^{4n-1}(p;q,\ldots,q)$. The bundles η_p, η_q over Y_p, Y_q induced by homotopy equivalences with the appropriate lens spaces have spherical Thom classes (see §2 in [3]), and by choice of the homotopy type of Y can both be lifted from some bundle η over Y. The covering projections define maps from S^{N+4n-1} to the Thom space Y^η with degrees p and q, so there exists a map of degree 1 and η has spherical Thom class.

It is interesting to observe that the argument of Theorem 3 shows that, at least for groups of order pq, the Spivak fibration can be reduced to an O_N-bundle, and hence that we may work in the smooth category.

In connection with step (c) above we pose the following

PROBLEM: which elements of $Wh(\pi)$ are geometrically realisable by simple homotopy equivalences $Y \to Y'$?

REFERENCES

1. M.F. Atiyah, Characters and cohomology of finite groups, Publ. Math. I.H.E.S. 9 (1961) 23-64.

2. J.W. Milnor, Groups which act on S^n without fixed points, Amer. J. Math. 79 (1957) 623-630.

3. M. Spivak, Spaces satisfying Poincaré duality, Topology 6 (1967) 77-101.

4. R.G. Swan, Periodic resolutions for finite groups, Annals of Math. 72 (1960) 267-291.

5. C.T.C. Wall, Surgery of compact manifolds, Liverpool University notes (1968).

6. J.A. Wolf, Spaces of constant curvature, McGraw-Hill Publishing Co. (New York) (1967).

University of Hull.

Offsetdruck: Julius Beltz, Weinheim/Bergstr.

MIX
Papier aus verantwortungsvollen Quellen
Paper from responsible sources
FSC® C105338

If you have any concerns about our products,
you can contact us on
ProductSafety@springernature.com

In case Publisher is established outside the EU,
the EU authorized representative is:
Springer Nature Customer Service Center GmbH
Europaplatz 3, 69115 Heidelberg, Germany

Printed by Libri Plureos GmbH
in Hamburg, Germany